endorsed by edexcel

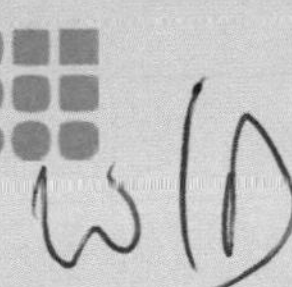

A Level Mathematics for Edexcel

Decision

D1

Brian Jefferson

OXFORD
UNIVERSITY PRESS

UNIVERSITY PRESS

Great Clarendon Street, Oxford OX2 6DP

Oxford University Press is a department of the University of Oxford.
It furthers the University's objective of excellence in research, scholarship,
and education by publishing worldwide in

Oxford New York

Auckland Cape Town Dar es Salaam Hong Kong Karachi
Kuala Lumpur Madrid Melbourne Mexico City Nairobi
New Delhi Shanghai Taipei Toronto

With offices in

Argentina Austria Brazil Chile Czech Republic France Greece
Guatemala Hungary Italy Japan South Korea Poland Portugal
Singapore Switzerland Thailand Turkey Ukraine Vietnam

First published 2008

British Library Cataloguing in Publication Data

Data available

ISBN 9780-19-911780 2

10 9 8 7 6 5 4 3

Printed in Great Britain by Ashford Colour Press Ltd, Gosport.

Paper used in the production of this book is a natural, recyclable product made from wood grown in sustainable forests. The manufacturing process conforms to the environmental regulations of the country of origin.

Acknowledgements

The photograph on the cover is reproduced courtesy of Paul Wilde/Adams Picture Library t/a apl

The publishers would like to thank the following for permission to reproduce photograpshs;
p28 Alamy; p54 iStockphoto; p70 Alamy; p82 Alamy; p92 Alamy; p106 Alamy; p124 iStockphoto; p154 iStockphoto.

The publishers would also like to thank Ian Bettison and Naz Amlani for their expert help in compiling this book.

About this book

Endorsed by Edexcel, this book is designed to help you achieve your best possible grade in Edexcel GCE Mathematics Decision 1 unit.

Each chapter starts with a list of objectives and an introduction. Chapters are structured into manageable sections, and there are certain features to look out for within each section:

Key points are highlighted in a blue panel.

Key words are highlighted in **bold blue** type.

Worked examples demonstrate the key skills and techniques you need to develop. These are shown in boxes and include prompts to guide you through the solutions.

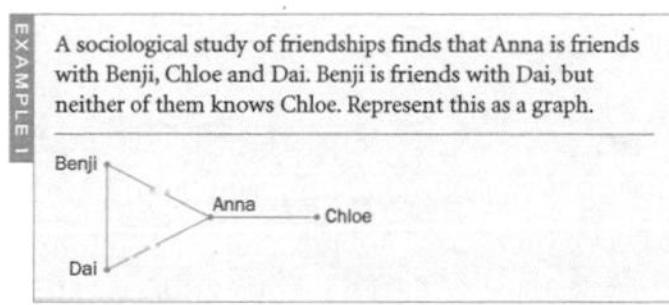

Derivations and additional information are shown in a panel.

Helpful hints are included as blue margin notes and sometimes as blue type within the main text.

Misconceptions are shown in the right margin to help you avoid making common mistakes.

Investigational hints prompt you to explore a concept further.

Each section includes an exercise with progressive questions, starting with basic practice and developing in difficulty. Some exercises also include 'stretch and challenge' questions marked with a stretch symbol .

At the end of each chapter there is a 'Review' section which includes exam style questions as well as past exam paper questions. There are also two 'Revision' sections per unit which contain questions spanning a range of topics to give you plenty of realistic exam practice.

The final page of each chapter gives a summary of the key points, fully cross-referenced to aid revision. Also, a 'Links' feature provides an engaging insight into how the mathematics you are studying is relevant to real life.

At the end of the book you will find full solutions, a key word glossary and an index.

Contents

D1

1 Algorithms

This chapter will show you how to

- understand the term algorithm
- follow an algorithm in the form of a list of instructions or a flowchart
- put items into groupings of restricted size using 'bin-packing' algorithms
- put lists into numerical or alphabetical order using sorting algorithms
- search for items in a list using the binary search algorithm.

Introduction

The term *Decision Mathematics* is used to cover a number of techniques for solving large-scale data processing and organisational problems.

Much of the underlying theory has been around for a long time, but its application only became possible with the development of computers. It now has an increasingly vital role to play in the world of industry and commerce.

This chapter will tell you about algorithms. Every technique in Decision Mathematics is built around a set of instructions – an algorithm – for solving a problem. You will become familiar with the idea of an algorithm, and learn about some specific examples.

You do not need knowledge of any particular mathematical topics in order to understand this chapter. You do however need to be able to show your working in a detailed, well-organised and logical way.

1.1 Algorithms

An algorithm is a well-defined sequence of steps leading to the solution of a problem of a given type.

Example 1 should give you a feel for the notion of an algorithm.

EXAMPLE 1

Find a procedure for completing the following simple puzzle. Six counters – three black and three white – are placed on a grid of seven squares, as shown.

The white counters may only move left to right, either sliding into a vacant space or jumping one black counter into a vacant space. The black counters move right to left under the same rules. The aim is to reverse the positions of the two sets of counters.

If you are not familiar with the solution to this puzzle, you might like to give it a try before reading on.

To state the solution to the puzzle, you could just list all fifteen moves needed. However, it is better to write a list of instructions so that the reader can decide on the right move at each stage.

One possible list is:

Step 1 Decide at random which colour is the 'active' colour
Step 2 Make the available active colour move
Step 3 If no further active colour move is possible, go to Step 6
Step 4 If the next available active colour move is a second slide move, do not make the move but go to Step 6
Step 5 Make the next available active colour move and go to Step 3
Step 6 If the puzzle is complete then stop
Step 7 Change the active colour and go to Step 2

You will not be expected to write algorithms from scratch, though you may be required to complete or modify an algorithm.

The instructions in Example 1 are more general than just the solution of the original puzzle. They work for any number of counters of each colour, even if the numbers of white counters and black counters are different. They form an algorithm, enabling anyone to solve all problems of this type.

You might like to try applying these instructions to another problem of this type.

An algorithm should

- enable someone to solve all problems of a particular type, just by following the instructions – no insight should be needed
- provide a clear 'next step' at each stage of the solution
- arrive at that solution in a finite and predictable number of steps
- ideally be well suited to computerisation.

The last point about computerisation is important because most real-world problems are too large to make a manual solution viable. However, the questions in this book and those in examinations are small in scale, so that you can solve them by hand in a reasonable time. As a result the solution may be obvious by inspection, but you must not be tempted to take short-cuts.

The examiner wants to know

- whether you know the correct algorithm to use
- whether you can follow the steps accurately to solve the problem.

You should always follow the correct steps and you must make your working clear.

Communicating an algorithm

An algorithm can be stated in two main ways.

1 A list of instructions.
  These instructions could sometimes be in the form of a computer program.

2 A flowchart.

You will not be expected to write a program.

This algorithm calculates the profit or loss of a transaction, given the cost price, C, and the selling price, S.

Here is the algorithm as a list of instructions:

Step 1 Input C, S
Step 2 Let $P = S - C$
Step 3 If $P \geqslant 0$ Then Print "Profit =", P: Go to Step 6
Step 4 Let $L = -P$
Step 5 Print "Loss =", L
Step 6 Stop

Here is the algorithm as a flowchart:

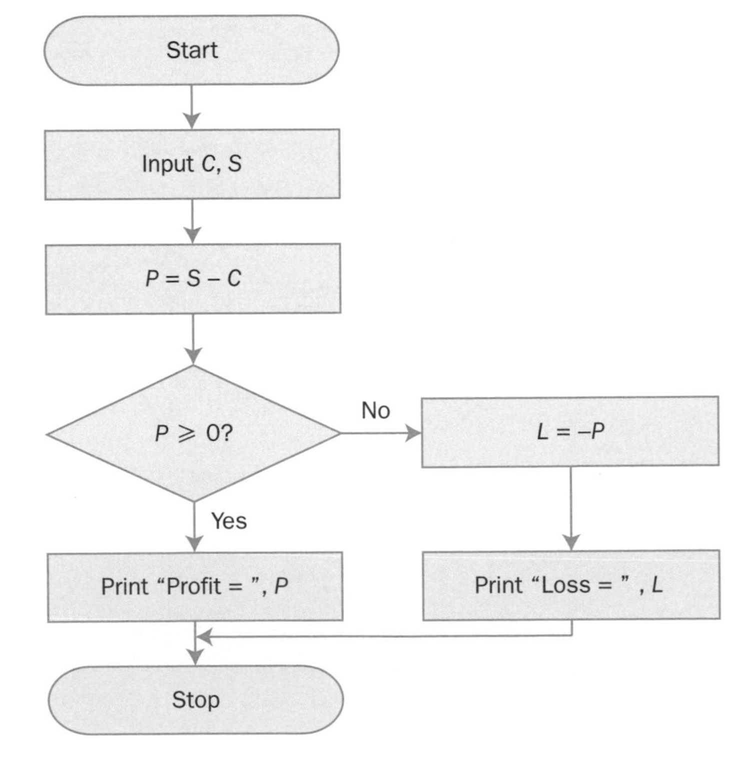

D1

EXAMPLE 2

This algorithm is designed to divide a positive integer, A, by another positive integer, B, giving a quotient, Q, and a remainder.

Step 1 Input A and B
Step 2 Let $Q = 0$
Step 3 If $A < B$ then go to Step 7
Step 4 Let $A = A - B$
Step 5 $Q = Q + 1$
Step 6 Go to Step 3
Step 7 Print 'Quotient = ', Q
Step 8 Print 'Remainder = ', A
Step 9 Stop

a Show the operation of the algorithm when $A = 100$ and $B = 23$.

b Display the algorithm in the form of a flowchart.

---

a The best way of showing the action of the algorithm is as a table of the values it produces. This is called a **trace table**.

Steps	A	B	Q	Notes
1,2	100	23	0	Set up the starting values
3,4,5	77	23	1	$A > B$ so subtract B from A, increase Q by 1
6,3,4,5	54	23	2	" " " "
6,3,4,5	31	23	3	" " " "
6,3,4,5	8	23	4	" " " "
3,7,8,9	8	23	4	$A < B$ so print 'Quotient = 4, Remainder = 8'

b

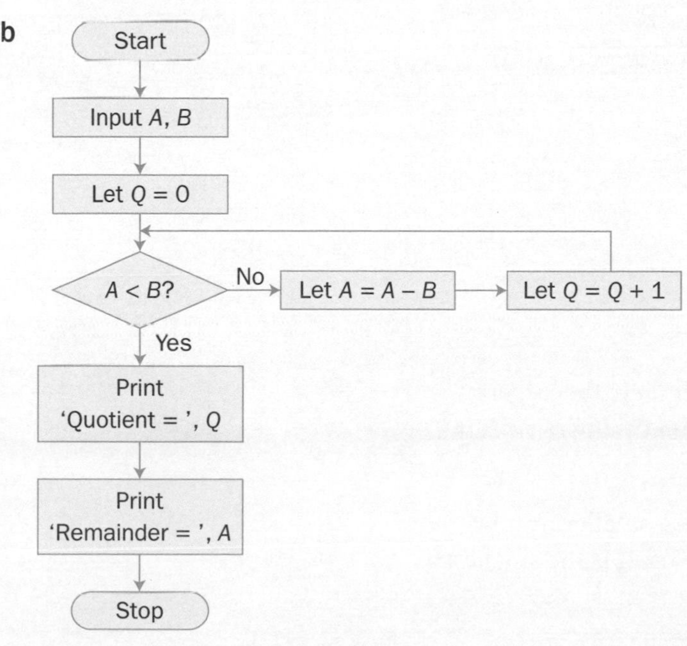

Follow this flowchart through and satisfy yourself that it correctly describes the algorithm.

Flowchart conventions

The main points to remember when drawing a flowchart are:

- the main sequence of instructions flows down the page and from left to right
- start/stop instructions use a box like this
- 'do this' instructions use a box like this
- yes/no decisions use a box like this 

EXAMPLE 3

For the algorithm in the flowchart

a apply the algorithm to the list $L(1) = 8$, $L(2) = 11$, $L(3) = 6$, $L(4) = 14$, $L(5) = 10$

b describe what the algorithm achieves

c state the effect of changing the decision '$M > L(I)$?' to '$M < L(I)$?'.

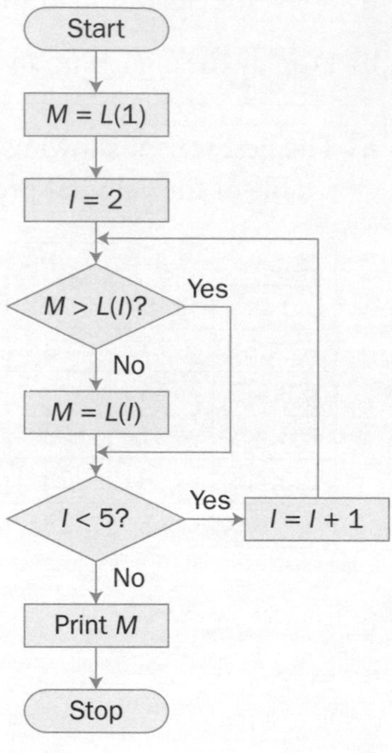

---

a The values taken by the variables M and I are

M	I
0	1
8	1
8	2
11	2
11	3
11	3
11	4
14	4
14	5
14	5

The variable I acts as a pointer to the position of successive values in the list.

The output value of $M = 14$.

b At each stage the algorithm remembers the larger of the two numbers. Hence the printed value of M is the largest number in the list.

c The algorithm now remembers the smaller number at each stage, so the output value of M is the smallest number in the list.

EXAMPLE 4

Susan wishes to generate a list consisting of the numbers 1 to 50 in a random order.
She proposes to use the following algorithm.

Step 1 Use the random number function on a calculator to obtain a random number between 1 and 50
Step 2 Write the number down unless it is already in the list
Step 3 If the list has 50 numbers, then stop
Step 4 Go to Step 1

Explain why this is not a viable algorithm.

An algorithm should end in a finite and predictable number of steps.

With the process described, as the list grows the chance of repeats increases. If Susan is unlucky she could go on for a very long time and there is a small probability that the number of steps could be bigger than any number you care to suggest.

Have a go at devising a better procedure for producing such a list.

D1

Exercise 1.1

1 List the output from this algorithm.

Step 1 Let $A = 1, B = 1$
Step 2 Print A, B
Step 3 Let $C = A + B$
Step 4 Print C
Step 5 Let $A = B, B = C$
Step 6 If $C < 50$ then go to Step 3
Step 7 Stop

2 **a** List the output from this algorithm.

Step 1 Let $A = 1, B = 1$
Step 2 Print A
Step 3 Let $A = A + 2B + 1$
Step 4 Let $B = B + 1$
Step 5 If $B \leqslant 10$ then go to Step 2
Step 6 Stop

b What does the algorithm achieve?

c What would be the effect of the following errors when typing this algorithm?
i Putting 'Print B' instead of 'Print A'.
ii Putting 'Let $A = A + B + 1$' instead of 'Let $A = A + 2B + 1$'
iii Putting '$B < 10$' instead of '$B \leqslant 10$'

3 **a** Apply this flowchart to

i $A = 12, B = 7$
ii $A = 53, B = 76$

b What does the algorithm achieve?

c The algorithm is very inefficient when, for example, $A = 211, B = 6$.

i Why is it inefficient?
ii How might you modify the flowchart to overcome this?

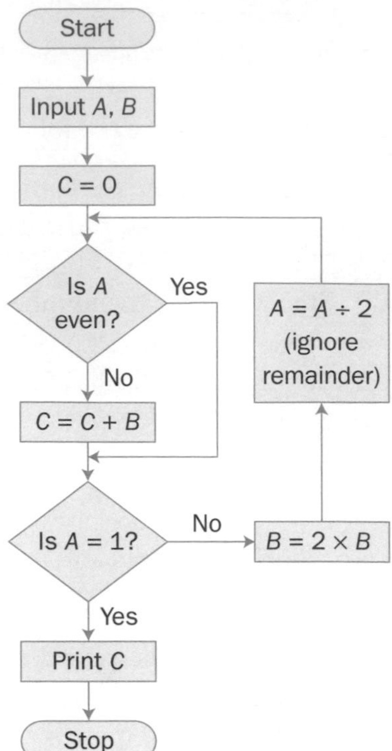

4 This is Euclid's algorithm. It finds the highest common factor of two numbers.

Step 1 Input A, B
Step 2 If $A \leqslant B$ then go to Step 4
Step 3 Swap A and B
Step 4 R = Remainder from $B \div A$
Step 5 If $R = 0$ then go to Step 9
Step 6 $B = A$
Step 7 $A = R$
Step 8 Go to Step 4
Step 9 Print 'HCF = ', A
Step 10 Stop

a Use the algorithm with starting values

i $A = 48, B = 132$
ii $A = 130, B = 78$

b Draw a flowchart to show this algorithm.

5 This flowchart is designed to find the nature of the roots of a quadratic equation $ax^2 + bx + c = 0$ by calculating the discriminant $d = b^2 - 4ac$.

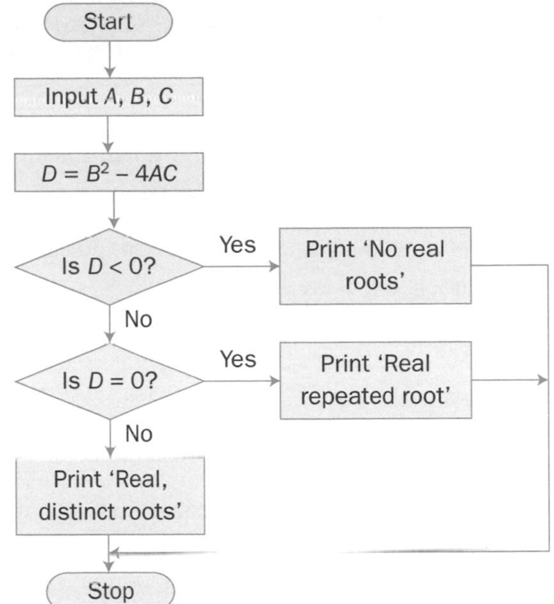

a Follow the flowchart for the equations

i $2x^2 - 6x - 3 = 0$

ii $3x^2 + 4x + 5 = 0$

iii $9x^2 - 6x + 1 = 0$

b The roots can be calculated, when they exist, using the quadratic formula

$$x = \frac{-b \pm \sqrt{b^2 - 4ac}}{2a}$$

Modify the flowchart so that it prints out the values of the roots where possible.

6 An examination consists of two papers, A and B, each marked out of 100. To gain a pass a candidate must score at least 50 on each paper, and have a total score of at least 120. To gain a distinction the total score must be at least 150. Describe, by means of a list of instructions or a flowchart, an algorithm for outputting 'Fail', 'Pass' or 'Distinction' for an input pair of marks.

7 A large group of adults travelling in a remote region find their route blocked by a river. They meet two children who own a boat, but the boat is only big enough to carry the two children or one adult. Nevertheless, the children manage to ferry the party across the river and return to their starting point.

a Give a list of instructions to explain, as economically as possible, how this was done.

b Find how many times the boat had to cross the river if there were ten adults in the party.

8 Draw a flowchart for an algorithm which prints out all the prime numbers between 100 and 1000.

9 Show that if the algorithm in Example 1 is applied to m white counters and n black counters, the puzzle will be completed in $(mn + m + n)$ moves.

1.2 Bin-packing algorithms

Some real-life problems require you to place items into compartments of a fixed size. Algorithms are used to find efficient solutions to these 'bin-packing' problems.

e.g. Loading vehicles onto a ferry with several lanes of equal length.

e.g. Suppose you wish to store items in bins 60 cm deep.
The heights, in cm, of the items are

8, 16, 12, 8, 45, 18, 30, 7, 10, 14, 9, 9, 52

The total of these heights is 238 cm, so you need at least 4 bins.
You want a packing order using the least number of bins.

Full-bin algorithm

One approach is to look for combinations of items which exactly fill a bin.

e.g. A possible solution is

Bin 1	8, 52	= 60
Bin 2	8, 7, 45	= 60
Bin 3	12, 18, 30	= 60
Bin 4	9, 9, 10, 14, 16	= 58

In this case you would need 4 bins. If the 45 and 52 had been 43 and 54, you would have neeed 5 bins even though the total was unchanged at 238 cm.

Full-bin algorithm:
1 search for combinations of values which exactly fill bins
2 for the remainder, place each item in the first available bin.

The problem with the full-bin algorithm is the trial-and-error nature of the first part. To do it fully you would have to examine all possible full-bin combinations to see which leads to the best solution. This is not practical for large problems and so other, more systematic procedures have been devised.

First-fit algorithm

This is essentially the second part of the full-bin algorithm.

In this case the first-fit algorithm gives you a worse result, needing 5 bins. However the first-fit algorithm provides a quick and systematic way of achieving what is usually a reasonably good solution, and can be used in large problems where the full-bin approach is not suitable.

First-fit algorithm:
taking the items in the order listed, place each item in the first bin which has room for it.

You may find it helpful to keep a record of the space left in the bins as they are filled.

8, ~~16~~, ~~12~~, 8, ~~45~~, ~~18~~, ~~30~~, ~~7~~, ~~10~~, ~~14~~, 9, 9, ~~52~~

	Contents of bin	Space left
Bin 1	8, 16, 12, 8, 7, 9	~~60~~, ~~52~~, ~~36~~, ~~24~~, ~~16~~, 9, 0
Bin 2	45, 10	~~60~~, ~~15~~, 5
Bin 3	18, 30, 9	~~60~~, ~~42~~, ~~12~~, 3
Bin 4	14	~~60~~, 46
Bin 5	52	~~60~~, 8

First-fit decreasing algorithm

This often gives a better result than the first-fit algorithm, but takes longer to process because of the need to sort the data.

First-fit decreasing algorithm:
1 sort the list of values into decreasing order of size
2 apply the first-fit algorithm.

None of these algorithms guarantees to give the **optimum** (best possible) result every time. They are **heuristic** algorithms, giving an acceptably good result in a reasonable length of time. No perfect algorithm exists (yet).

Applying this to the example, you get

~~52~~, ~~45~~, ~~30~~, ~~18~~, ~~16~~, ~~14~~, ~~12~~, ~~10~~, ~~9~~, ~~9~~, ~~8~~, ~~8~~, ~~7~~

	Contents of bin	Space left
Bin 1	52, 8	~~60~~, ~~8~~, 0
Bin 2	45, 14	~~60~~, ~~15~~, 1
Bin 3	30, 18, 12	~~60~~, ~~30~~, ~~12~~, 0
Bin 4	16, 10, 9, 9, 8, 7	~~60~~, ~~44~~, ~~34~~, ~~25~~, ~~16~~, ~~8~~, 1

EXAMPLE 1

A carpenter wants to cut 12 pieces of wood – two 40 cm long, four 60 cm long, three 80 cm long, one 120 cm long and two 140 cm long. The wood is sold in 2.4 m lengths.

a What is the minimum number of lengths required?

This is called the **lower bound** for the number of lengths required.

b Use the first-fit decreasing algorithm to suggest a cutting plan.

c Show that there exists a better solution to the problem.

a The total of the required pieces is 9.6 m, so at least 4 lengths will be needed.

b Sort the data into decreasing order and apply the first-fit procedure:

~~140~~, ~~140~~, ~~120~~, ~~80~~, ~~80~~, ~~80~~, ~~60~~, ~~60~~, ~~60~~, ~~60~~, ~~40~~, ~~40~~

	Contents of bin	Space left
Length 1	140, 80	~~240~~, ~~100~~, 20
Length 2	140, 80	~~240~~, ~~100~~, 20
Length 3	120, 80, 40	~~240~~, ~~120~~, ~~40~~, 0
Length 4	60, 60, 60, 60	~~240~~, ~~180~~, ~~120~~, ~~60~~, 0
Length 5	40	~~240~~, 200

This cutting plan requires 5 lengths and involves wastage totalling 2.4 m.

c By inspection you can see that a more efficient solution would be

Length 1 140, 60, 40 = 240
Length 2 140, 60, 40 = 240
Length 3 80, 80, 80 = 240
Length 4 120, 60, 60 = 240

You might have found this solution using the full-bin algorithm, but it would depend on the order in which you chose the full bins. e.g. Noticing immediately that the four 60 cm items 'fill a bin' would not have arrived at the best result. This shows why the full-bin approach is unsatisfactory.

Exercise 1.2

1 Twelve items are to be packed in bins of height 18 cm. The heights of the items, in cm, are

5, 3, 9, 6, 11, 2, 3, 7, 6, 4, 6, 10

a Calculate the lower bound for the number of bins involved.

b Use the first-fit algorithm to obtain a possible solution.

c Use the first-fit decreasing algorithm to obtain an improved solution.

2 Nineteen schools send a total of 100 students to an A-level revision day. The numbers from the schools are

5, 3, 3, 4, 6, 4, 6, 8, 10, 6, 5, 6, 2, 6, 4, 2, 8, 2, 10

The students are to be placed in workshop groups, each containing a maximum of 20 students. Students from the same school are to be in the same group.

a Use a full-bin approach to show that the students can fit into five workshops.

b Show that the first-fit algorithm does not give an optimum result.

c Apply the first-fit decreasing algorithm and show that it gives an optimum allocation.

3 A transport company is employed to move fourteen items whose masses (in tonnes) are

3.0, 1.2, 3.5, 2.5, 0.5, 1.0, 1.0, 2.5, 1.2, 0.7, 2.0, 1.8, 2.1, 1.2

The available van has a maximum payload of 5 tonnes.

a Use the first-fit algorithm to devise a possible loading plan. How many trips will be needed?

b Use the first-fit decreasing algorithm to obtain an improved plan. How many trips are needed with this plan?

c Explain why it is not possible to do the job in fewer trips than this.

4 An upholsterer needs to cut various lengths of material from standard 12 m rolls. The lengths required, in metres, are

2, 2, 3, 3, 4, 6, 7, 9

a Show that neither the first-fit nor the first-fit decreasing algorithm gives a solution using less than four rolls.

b Show that it is possible to solve the problem using only three rolls.

5 A music fan stores CDs in boxes with a capacity of 15 discs. She classifies the CDs according to the artiste, and intends to ensure that all CDs by a given artiste are stored in the same box. The numbers of CDs in each classification are

6, 11, 4, 12, 9, 3, 5, 5, 6, 6, 2, 8, 9

Using a suitable algorithm and showing all your working, investigate whether she can store her CDs in six boxes.

6 A company has a number of tasks to be completed in a 35-hour week. Each task is to be done from start to finish by one person. The projected lengths of the tasks are shown in the table. The company plans to hire temporary staff for these tasks.

No of hours	4	5	6	7	8	9
No of tasks	8	10	4	6	3	4

a Show that a task allocation produced using the first-fit decreasing algorithm would require the company to hire seven staff.

b Show that by modifying the allocation in part **a** the company need only hire six staff.

7 The organiser of a local talent contest wishes to arrange the evening as three 45-minute sections with drink and chat intervals between. There are ten acts in the competition with running times as shown (in minutes).

6, 8, 10, 20, 6, 10, 24, 15, 20, 15

a Show that the first-fit algorithm fails to give a suitable running order.

b Use the first-fit decreasing algorithm to find a running order.

c The organiser decides that there should be at least three acts in each section. Modify your answer to part **b** to allow for this.

1.3 Sorting algorithms

One of the most common data processing tasks is sorting an unordered list into numerical or alphabetical order. For long lists and for computer implementation this requires an algorithm.

Short lists, such as those in exam questions, can be easily sorted 'by inspection'. Remember that the question is testing whether you know the algorithm, not whether you can sort the list.

Sorting a long list can take a lot of processing time. Many sorting algorithms have been devised. You need to know the details of two – the bubble sort and the quick sort.

If you search the internet you will find twenty or more sorting algorithms. They vary enormously in speed, and there is no single best method.

Bubble sort

Bubble sort

- **First** pass
  Compare the 1st and 2nd numbers. Swap them if in the wrong order.

  Compare/swap the 2nd and 3rd numbers, then the 3rd and 4th numbers, and so on to the end of the list.
- **Subsequent passes**
  Repeat the pass process, leaving out the final number from the previous pass.
- **Terminating**
  Finish when either the last pass had only two numbers
  or a complete pass produced no swaps.

e.g. Sorting the list 5, 3, 6, 2 into ascending order, the first pass looks like this:

5 3 6 2 Compare first and second. Swap.

3 5 6 2 Compare second and third. No swap needed.

3 5 6 2 Compare third and fourth. Swap.

3 5 2 6 The 6 is now in the correct place.

Because the final number is now in the right place, you can ignore it on the second pass.

3 5 2 | 6 Compare first and second. No swap needed.

3 5 2 | 6 Compare second and third. Swap.

3 2 5 | 6 The 5 is now in the correct place.

On the next pass you can ignore the final two numbers.

3 2 | 5 6 Compare first and second. Swap.

2 3 | 5 6 The list is now sorted.

If at any stage you make a complete pass without doing any swaps, it means the list is in the correct order, so you stop.

EXAMPLE 1

Arrange the numbers 4, 8, 2, 6, 3, 5 in ascending order using the bubble sort algorithm. Record the number of comparisons and the number of swaps made at each stage.

First pass

4	8	2	6	3	5
4	8 ◄──► 2		6	3	5
4	2	8 ◄──► 6		3	5
4	2	6	8 ◄──► 3		5
4	2	6	3	8 ◄──► 5	
4	2	6	3	5	8

In the tables, the shading shows the comparison being made at each stage, with the arrow indicating if a swap is necessary.

The 8 has moved to its correct place after this pass.

The first pass involved 5 comparisons and 4 swaps.

Second pass

4 ◄──► 2		6	3	5	8
2	4	6	3	5	8
2	4	6 ◄──► 3		5	8
2	4	3	6 ◄──► 5		8
2	4	3	5	6	8

The 6 has moved to its correct place after this pass.

The second pass involved 4 comparisons and 3 swaps.

Third pass

2	4	3	5	6	8
2	4 ◄──► 3		5	6	8
2	3	4	5	6	8
2	3	4	5	6	8

The 5 has moved to its correct place after this pass.

The third pass involved 3 comparisons and 1 swap.

Fourth pass

2	3	4	5	6	8
2	3	4	5	6	8
2	3	4	5	6	8

There were no swaps on this pass, so the sort is complete.

The fourth pass involved 2 comparisons and 0 swaps.

Different sorting algorithms vary in how efficiently they accomplish the task. This is measured by how many operations are involved, which in the case of the bubble sort is the number of comparisons and swaps made.

If a computer is being used, this translates into the amount of processing time needed.

In Example 1 the bubble sort involved a total of 14 comparisons and 8 swaps.

Quick sort

This algorithm takes a different approach from the comparison/swap method of the bubble sort.

The quick sort can be defined in compare/swap terms, but this lies outside the present syllabus.

One number is chosen to be the **pivot**. The rest are split into two sub-lists – numbers less than the pivot and numbers greater than the pivot. The procedure is then repeated for each sub-list, with its own pivot, until the list is sorted.

If a number is equal to the pivot, you could put it in either sub-list. In this book the 'equal' numbers are included in the 'less than' sub-list.

The list starts in random order, so the choice of pivot is arbitrary. Some authors use the first value in the list, but this syllabus will expect you to use the middle value in the list. If there are two 'middle' values, either can be used – this book will use the second one as the pivot.

This is in line with the practice used in binary search – see Section 1.4.

Quick sort

Step 1 Choose the middle number as the pivot.
Step 2 For each remaining number in the list:
if number $\leqslant$ pivot, place number in sub-list before pivot; otherwise place number in sub-list after pivot.
Step 3 Quick sort sub-lists containing two or more numbers

Numbers go in the sub-lists in the order in which they appeared in the original list. (This is an arbitrary choice, but an algorithm must be definite at each stage.)

The quick sort is an example of a **recursive algorithm**, that is, one which contains itself.

EXAMPLE 2

Arrange the numbers 4, 8, 3, 1, 7, 5, 12, 13, 2, 6 in ascending order using the quick sort algorithm.

4	8	3	1	7	5	12	13	2	6

Taking the 5 as the pivot gives these sub-lists.

The 5 is the second of the two middle numbers.

4	3	1	2	5	8	7	12	13	6

The pivots for the two sub-lists are 1 and 12, giving

The 1 is the second of the two middle numbers.

1	4	3	2	5	8	7	6	12	13

Two sub-lists have two or more numbers. The pivots are 3 and 7.

1	2	3	4	5	6	7	8	12	13

The remaining sub-lists each have length 1, so the process is complete.

To bubble sort this list you would have made 45 comparisons, whereas the quick sort involved just 20.

The quick sort is usually more efficient than the bubble sort.

Exercise 1.3

1 Use the bubble sort algorithm to sort this list into

a ascending order **b** descending order.

12 4 16 5 9 2 4

Show the result of each pass.

2 Repeat question **1** using the quick sort algorithm.

3 Sort this list into ascending order using the bubble sort algorithm.

22 26 14 20 12 9 11 15 10

4 Sort this list into ascending order using the quick sort algorithm.

9 17 6 19 16 13 7 17 12 9

5 A class has students with surnames

Harris, Thomas, Patel, Frobisher, Cheung, Allen and Lee.

Sort these into alphabetical order using

a bubble sort **b** quick sort.

6 Another sorting algorithm is the interchange sort. For ascending order the algorithm is:

Step 1 Find the largest number. Swap it (if necessary) with the last number

Step 2 Ignoring the last number, repeat Step 1 with the remaining list

Step 3 Repeat Step 2 until there is only one number in the remaining list

a Use the interchange sort to arrange the list 1, 5, 9, 3, 11, 7, 13 in ascending order. Record the state of the list after each step, and count the number of comparisons and swaps needed.

b Sort the same list using the bubble sort. Which method was more efficient?

7 The worst-case scenario for a bubble sort is that the original list is in the reverse of the required order.

a For this worst case, find the number of comparisons and swaps when sorting a list of five numbers.

b How many would be needed for a list of
i 10 **ii** 20 **iii** n numbers?

8 The worst-case scenario for a quick sort depends on the convention being used for choosing the pivot. For a quick sort using the middle number convention

a give a worst case starting order for the numbers 1, 2, 3, 4, 5, 6, 7

b find the number of comparisons (between numbers and the pivot) which would be needed in this case.

1.4 Searching

When you ask a computer to search a database for an item (which may or may not be there), the computer does not check every entry in the file. It uses an algorithm to reduce the number of items it needs to examine.

Here is an illustration of the most common search algorithm:

Your friend chooses a number between 1 and 20, and asks you to identify it by asking as few questions as possible.

You could just keep guessing different numbers until you hit on the right one. However, unless you strike lucky, the size of the problem only reduces by one at each attempt. If you were unlucky, you could take 19 questions.

The best method is to ask how the number compares (equal, less or greater) with the middle number of the list. Either the number is found or the problem size is halved.

As in Section 1.3, if there are two 'middle' values choose the second of these.

Suppose your friend chooses 14. The middle number is 11.

11 is the second of the two middle numbers.

1	2	3	4	5	6	7	8	9	10	11	12	13	14	15	16	17	18	19	20

Your friend tells you 'The number is above 11'.
The number lies between 12 and 20. The middle number is 16.

12	13	14	15	16	17	18	19	20

Your friend tells you 'The number is below 16'.

The number lies between 12 and 15. The middle number is 14.

14 is the second of the two middle numbers.

12	13	14	15

Your friend tells you 'The number is equal to 14'.

You have found the number in three attempts.

The most attempts you would have needed is four (e.g. if your friend had chosen 13).

This process is called a binary search. It is widely used for searching sorted lists of data.

The list to be searched must be sorted into order (numerical or alphabetical) before a binary search can be done.

Binary search

To search a list for the item '*TARGET*'

Step 1 Find middle item, M, of search list.

Step 2 If $TARGET = M$ then $TARGET$ has been found: Stop.

Step 3 If $TARGET < M$ then new search list is items below M.
If $TARGET > M$ then new search list is items above M.

Step 4 Repeat from Step 1 until $TARGET$ is found or search list is empty ($TARGET$ not in list).

The difference between a general binary search and the number guessing game is that the target item may not appear in the list.

You need to be able to find the middle position of a search list.

For a search list starting at position n_1 and ending at position n_2:

if the number of items is odd, the middle position is $m = \frac{1}{2}(n_1 + n_2)$

if the number of items is even, the middle position is the next integer value above this (the second of the two middle items).

Notation: $\left\lceil \frac{1}{2}(n_1 + n_2) \right\rceil$ means 'the smallest integer greater than or equal to $\frac{1}{2}(n_1 + n_2)$'.

The middle position of a search list from position 12 to position 20 is $\left\lceil \frac{1}{2}(12 + 20) \right\rceil = 16$.

The middle position of a search list from position 15 to position 24 is $\left\lceil \frac{1}{2}(15 + 24) \right\rceil = \left\lceil 19\frac{1}{2} \right\rceil = 20$.

EXAMPLE 1

Perform a binary search on the following list to look for

a Earle **b** Underwood.

1. Asquith	2. Auden	3. Berisford	4. Dubarry
5. Earle	6. Ford	7. Harris	8. Khan
9. Martin	10. O'Malley	11. Treacher	12. Wilson
13. Zuckerman			

---

a The middle position in the list is $\left[\frac{1}{2}(1+13)\right] = 7$, which is Harris.
Earle comes before Harris, so the new search list is

1. Asquith	2. Auden	3. Berisford
4. Dubarry	5. Earle	6. Ford

The middle position of this list is $\left[\frac{1}{2}(1+6)\right] = \left[3\frac{1}{2}\right] = 4$, which is Dubarry.
Earle comes after Dubarry, so the new search list is
5. Earle 6. Ford

The middle position of this list is $\left[\frac{1}{2}(5+6)\right] = \left[5\frac{1}{2}\right] = 6$, which is Ford.
Earle comes before Ford, so the new search list is just
5. Earle

The middle of this list is obviously position 5, which is Earle. The algorithm has located Earle in position 5.

b The middle position in the list is $\left[\frac{1}{2}(1+13)\right] = 7$, which is Harris.
Underwood comes after Harris, so the new search list is

8. Khan	9. Martin	10. O'Malley
11. Treacher	12. Wilson	13. Zuckerman

The middle position of this list is $\left[\frac{1}{2}(8+13)\right] = \left[10\frac{1}{2}\right] = 11$, which is Treacher.
Underwood comes after Treacher, so the new search list is
12. Wilson 13. Zuckerman

The middle position of this list is $\left[\frac{1}{2}(12+13)\right] = \left[12\frac{1}{2}\right] = 13$, which is Zuckerman.
Underwood comes before Zuckerman, so the new search list is just 12. Wilson

The middle of this list is obviously position 12, which is Wilson.
Underwood comes before Wilson, but the list is now empty, so Underwood is not on the list.

Remember that it is your understanding of the binary search process that is being tested. It is important to show details of your working.

Exercise 1.4

For binary search questions, your working should show the sequence of items examined during the search.

1 Identify the middle item in these lists.

a 1 – Aktar
2 – Amner
3 – Charles
4 – Ferriday
5 – Garton
6 – Nish
7 – Patel
8 – Quedgeley

b 9 – Dewsbury
10 – Evesham
11 – Exeter
12 – Garstang
13 – Holmfirth
14 – Jarrow

c 16 – Fiat
17 – Ford
18 – Hyundai
19 – Jaguar
20 – Kia
21 – Lada
22 – Lotus
23 – Nissan
24 – Seat
25 – Skoda
26 – Toyota

2 Use binary search to look for

a Nish in the list from question **1a**

b Exhall in the list from question **1b**

c Jaguar in the list from question **1c**.

3 The table shows the names of children booked for a coach trip.

No	Name	No	Name	No	Name	No	Name
1	Barry	6	Edith	11	Laurence	16	Quentin
2	Boris	7	Floella	12	Mabel	17	Rebecca
3	Catherine	8	Gerald	13	Nuria	18	Tristan
4	Cedric	9	Ingrid	14	Omar	19	Xavier
5	Declan	10	Juan	15	Petula	20	Zoe

Use binary search to look for

a Omar b Floella c Mary.

D1

4 A company's database is set up so that a customer's record can be found by entering their postcode. The table shows the list. Use binary search to find the name of the customer whose postcode is

a BS7 8NB **b** TA6 8KC.

No	Postcode	Name	No	Postcode	Name	No	Postcode	Name
1	BA4 6AS	L. Muswell	11	BS6 2JG	S. Voce	21	TA1 5HA	E. O'Flynn
2	BA4 7TS	K. Bennett	12	BS7 3LE	O. Banwell	22	TA3 8FF	U. Paine
3	BA4 7VA	A. Thorpe	13	BS7 8NB	I. R. Smith	23	TA3 9BS	T. Dearing
4	BA5 3HG	B. Fitzroy	14	BS9 6BS	F. Button	24	TA4 4AT	B. I. Twist
5	BA6 9FR	G. Mander	15	BS12 2IJ	D. Aire	25	TA4 7NS	V. Tingle
6	BA11 2JJ	R. Singh	16	BS17 3YY	W. Jones	26	TA5 2MN	D. Harold
7	BA12 1PJ	I. West	17	BS21 9AB	J. Kryzwicki	27	TA6 8KC	S. Elliott
8	BA22 7JP	P. D. Quicke	18	BS21 9CE	P. Gomez	28	TA6 8KT	C. Pericles
9	BS3 4HR	N. Ambrose	19	TA1 4EF	H. Smith-Ball	29	TA8 3DD	Y. Nott
10	BS3 4HW	B. Fouracres	20	TA1 5GK	A. McKinley	30	TA8 4YD	R. Taylor

5 A market research company plans to send interviewers to a random sample of 40 houses in a city street of 300 houses. The owner of number 227 contacts them to ask if her house is involved. Perform a binary search of this list of house numbers to answer her query.

Interview no.	1	2	3	4	5	6	7	8	9	10
House no.	**14**	**55**	**64**	**69**	**70**	**71**	**83**	**120**	**121**	**129**
Interview no.	11	12	13	14	15	16	17	18	19	20
House no.	**141**	**143**	**145**	**152**	**154**	**163**	**166**	**183**	**186**	**188**
Interview no.	21	22	23	24	25	26	27	28	29	30
House no.	**189**	**192**	**198**	**200**	**202**	**203**	**204**	**209**	**220**	**229**
Interview no.	31	32	33	34	35	36	37	38	39	40
House no.	**231**	**232**	**234**	**241**	**252**	**274**	**278**	**280**	**285**	**287**

6 A game consists of one player choosing a square on the grid shown. The other player then guesses a square and is told whether the target square is left or right and up or down from the guess.
In a particular game the target square (T, as shown) is M9.
Give a list of guesses, based on the binary search algorithm, to identify the target.

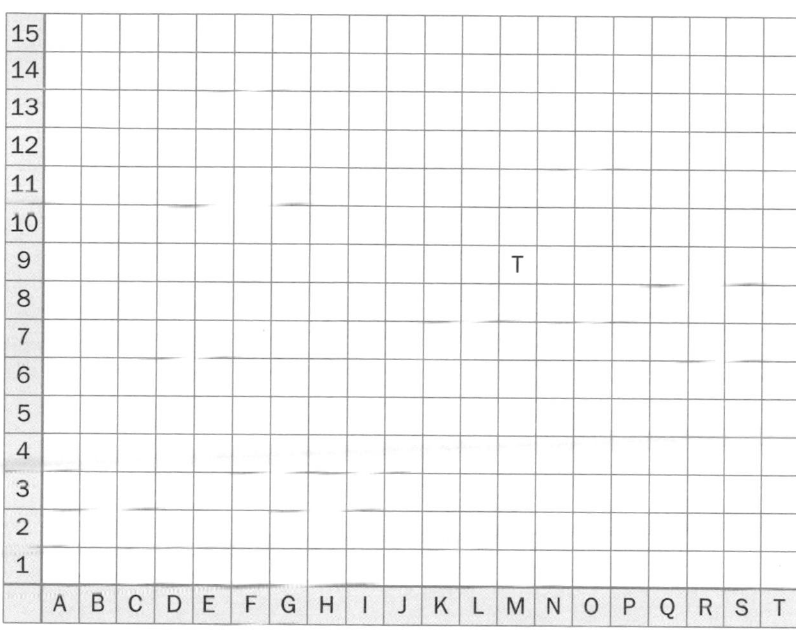

7 The number of records that must be examined to locate a particular item (or show that it is not in the list) depends on the number of items in the list.

a For a binary search, find the maximum number of records that must be examined for a list of

i 3 items ii 4 items iii 7 items
iv 8 items v 15 items.

b Generalise your results from part **a** to a list of n items.

c A linear search involves examining each record in turn to find the item sought. Repeat parts **a** and **b** for a linear search.

d Investigate the mean number of records examined using
i linear search ii binary search.

Assume in part **d** that the item is in the list.

D1

Review 1

1 a List the output from this algorithm.

Step 1 $A = 0, B = 1, C = 0$
Step 2 $A = A + B$
Step 3 Print A
Step 4 $C = C + 6$
Step 5 $B = B + C$
Step 6 If $A < 500$ then go to Step 2
Step 7 Stop

b What does the algorithm achieve?

2 a Execute the algorithm shown in the flowchart for each of these input values.

i 12 **ii** 8 **iii** 30 **iv** 32

In each case list the values taken by A and C as the algorithm proceeds, together with the output.

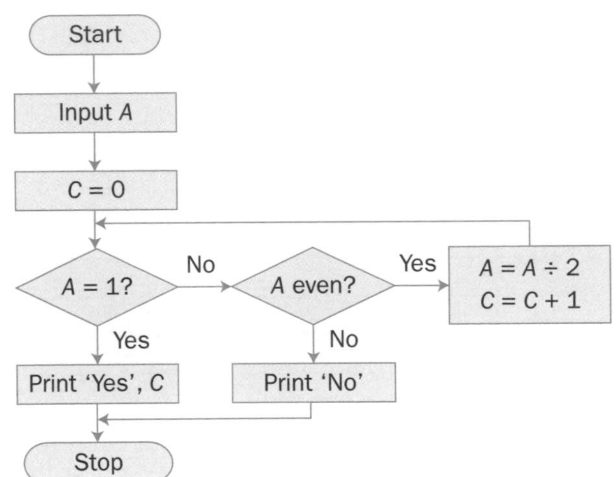

b Explain what the algorithm achieves.

3 a What would be the output from this algorithm?

Step 1 $A = 1, B = 1$
Step 2 Print A, B
Step 3 $C = A + B$
Step 4 Print C
Step 5 $A = B, B = C$
Step 6 If $C < 50$ go to Step 3
Step 7 Stop

b Draw a flowchart to show the algorithm from part **a**.

4 Use the quick sort algorithm to sort these children into increasing order of age.

Name	Chloe	Adam	Dan	Sally	Miguel	Karl	Tanya	Usha	Pete	Fay
Age (years)	8	5	15	7	13	9	12	6	9	10

5 A ferry company has a small vehicle ferry with four lanes, each 40 m long. The spaces in metres needed for the vehicles on a certain day (in order of arrival) were

4, 6, 5, 12, 14, 5, 5, 6, 4, 4, 14, 16, 10, 5, 6, 5, 8, 13, 6, 4, 4

a Use the first-fit algorithm to place them in the lanes and record what you find.

b It is then suggested that the approach is amended to

As vehicles arrive, line them up in two queues, one for vehicles 10 m or over, the other for shorter vehicles. Load first the long vehicles, then the short vehicles, using the first-fit algorithm.

What would be the result of using this modified approach for the vehicles listed above?

6 A cable car has space for 15 passengers. Passengers arrive in travel groups. At the start of the day there is a long queue, with the number of people in each travel group as shown.

9 4 3 8 6 5 9 6 5 4 8 4 3

The members of each travel group wish to stay together.

a Find the lower bound for the number of trips that might be required to clear the backlog.

b Show that the first-fit decreasing algorithm does not give a solution involving this minimum number of trips.

c Find by inspection a plan which does the job in the minimum number of trips.

7 A plumber buys copper pipe in 4 m lengths. For a particular job he requires the lengths shown.

Length of pipe (m)	0.3	0.4	0.5	0.6	0.7	1.2	1.5	2.0
Number required	2	3	1	2	3	2	1	3

Devise a cutting plan to do the job with as little waste as possible.

8 Use the bubble sort algorithm to sort this list into ascending order. Show the state of the list after each pass.

12 4 9 15 11 12 8 2 7

9 Repeat question 7 using the quick sort algorithm, again showing the list after each stage.

10 Use the bubble sort algorithm to sort this list of surnames into alphabetical order.

Hughes, Benn, Carter, Lee, Jenner, Afsar, Brown, Burns, Wilson, Bligh

11 Another method of sorting is using the shuttle sort algorithm:
First pass
Compare the 1st and 2nd numbers. Swap them if they are in the wrong order.
Second pass
Compare the 2nd and 3rd numbers. Swap them if they are in the wrong order.
If they were swapped, redo the first pass.
Third pass
Compare the 3rd and 4th numbers. Swap them if they are in the wrong order. If they were swapped, redo the second pass.
Subsequent passes
Repeat the above process, introducing the next number in the list at each stage. You backtrack through the list, comparing and swapping until no swap occurs.
Terminating
A list of n items terminates after $(n - 1)$ passes.

a Use the shuttle sort algorithm to sort this list into ascending order. Find the total number of comparisons and swaps made.

125 86 315 97 130 266 45 120 284 90

b Sort the same list using the bubble sort.
Find the total number of comparisons and swaps made.

12 This list of numbers is to be sorted into descending order.

55 80 25 84 25 34 17 75 3 5

a Perform a bubble sort to obtain the sorted list, giving the state of the list after each complete pass.

The numbers in the list represent weights, in grams, of objects which are to be packed into bins that hold up to 100 g.

b Determine the least number of bins needed.

c Use the first-fit decreasing algorithm to fit the objects into bins which hold up to 100 g. [(c) Edexcel Limited 2002]

D1

13 Perform a binary search to look for each of the following names in the list:

a Simpson **b** Harman **c** Timms

1 – Berisford	5 – Harris	9 – Mountjoy	13 – Trueman
2 – Cheung	6 – Ingoe	10 – Ng	14 – Unwin
3 – Faruq	7 – James	11 – Simpson	15 – Wyndham
4 – Harman	8 – MacAllister	12 – Thompson	16 – Young

14 An ornithologist makes a list of all the species seen during a bird watching trip.

No	Species	No	Species	No	Species	No	Species	No	Species
1	Blue tit	6	Egret	11	Jay	16	Pied wagtail	21	Swallow
2	Brambling	7	Goldfinch	12	Kingfisher	17	Robin	22	Whitethroat
3	Bullfinch	8	Great tit	13	Lapwing	18	Song thrush	23	Widgeon
4	Buzzard	9	Greenfinch	14	Nuthatch	19	Sparrowhawk	24	Wren
5	Dunnock	10	Heron	15	Pheasant	20	Starling	25	Yellowhammer

a Use the binary search algorithm to look for these birds in the list:

i Heron ii Nuthatch iii Swift

In each case state the number of records which had to be examined.

b A linear search involves examining every record in turn until the required record is found. How many records would have been examined if this technique had been used instead of binary search in part a?

15

1. Glasgow	6. Birmingham
2. Newcastle	7. Cardiff
3. Manchester	8. Exeter
4. York	9. Southampton
5. Leicester	10. Plymouth

A binary search is to be performed on this list to locate the name Newcastle.

a Explain why a binary search cannot be performed with the list in its present form.

b Using an appropriate algorithm, alter the list so that a binary search can be performed. State the name of the algorithm you use.

c Use the binary search algorithm on your new list to locate the name Newcastle.

1 Exit

Summary

Refer to

- An algorithm is a set of instructions, described by a list or a flowchart, for solving all problems of a given type. 1.1
  It should
  - provide a clear 'next step' at each stage of the solution
  - arrive at that solution in a finite and predictable number of steps
  - ideally be well suited to computerisation.
- Bin-packing algorithms – full-bin, first-fit, first-fit decreasing – are used to allocate items to groups of a fixed size. 1.2
- Sorting algorithms – bubble sort, quick sort – are used to sort lists of items into numerical or alphabetical order. 1.3
- The binary search algorithm is used to search an ordered list for a particular item. 1.4

Links

This chapter has included several examples of how sorting algorithms can be used to solve one-dimensional problems in real life.

The bin-packing algorithm has many applications such as scheduling lectures of different lengths into a given number of lecture theatres or compressing data files on a hard disk into clusters of a fixed size.

There is still much research ongoing into algorithms for two- and three-dimensional bin packing problems. Most algorithms rely to some extent on the first-fit principle.

An example of a two-dimensional problem is cutting rectangles of different sizes from standard sheets of metal, with the aim of minimising wastage. Three-dimensional problems include loading pallets onto lorries or packages into shipping containers. Such problems may have additional constraints to do with the weight and balance of the load.

2 Graphs and networks

This chapter will show you how to
- use the standard terminology of graph theory
- represent a graph or network using edge and vertex sets, or using adjacency or distance matrices
- use graphs and networks to model problems
- determine whether a graph is traversable.

Introduction

You can illustrate and analyse many problems in Decision Mathematics using a network of points and lines. This chapter is about the mathematics of such diagrams, which are known as graphs.

Many special words are used for types of graphs and for particular features of graphs. It is important that you become familiar with this vocabulary.

2.1 Graph theory – definitions

The term graph is used for a diagram involving a set of points and interconnecting lines. Each point is called a vertex (plural vertices) and a line joining two points is called an edge.
Some people use node instead of vertex and arc instead of edge.

The present syllabus uses **both** sets of terms interchangeably.

> A graph consists of a number of points (vertices or nodes) connected by a number of lines (edges or arcs).
>
> Which vertex is connected to which is important.
> The shape or layout of the diagram is irrelevant.

A tidy layout helps you to visualise the situation, but it is not mathematically important.

These three diagrams show the same set of vertices and connections – they are effectively the same graph.

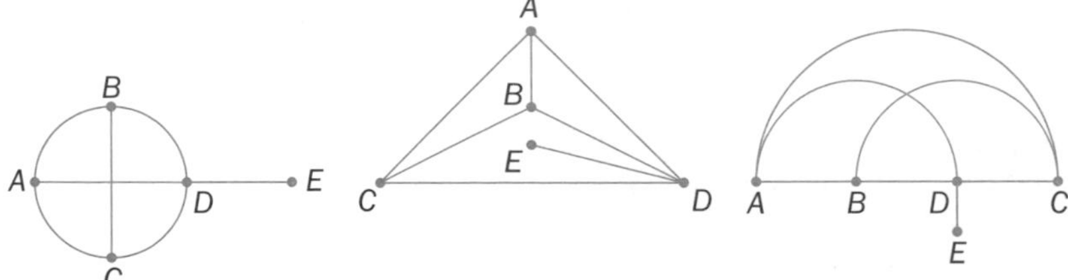

Check that you can see how these are essentially the same graph.

Graphs like these are called isomorphic ('of the same form').

A graph is connected if it is possible to travel from any vertex to any other vertex (perhaps passing through others on the way).

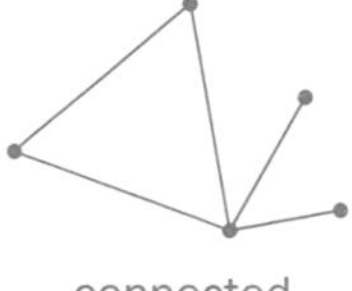

connected

disconnected

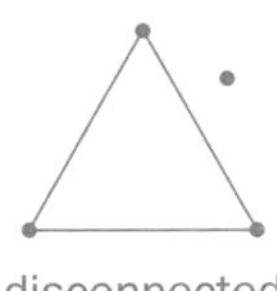

disconnected

Two or more edges may connect the same pair of vertices.
There may be a loop connecting a vertex to itself.

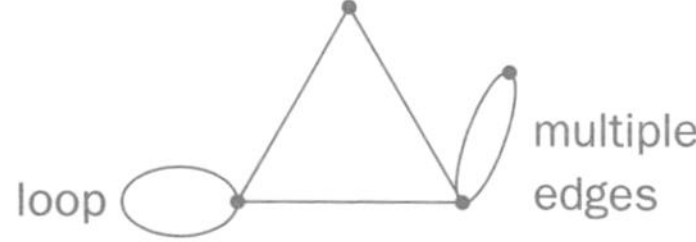

A graph with no loops or multiple edges is a simple graph.

The graph formed by some of the vertices and edges of a given graph is a subgraph of the given graph.

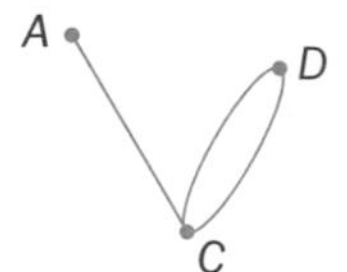

is a subgraph of

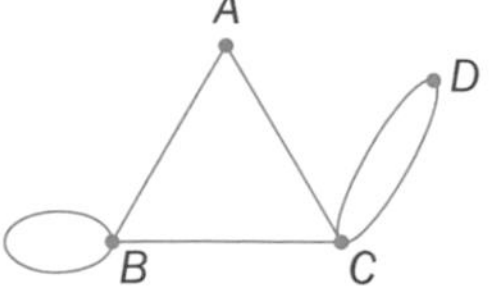

D1

If a simple graph has an edge connecting each possible pair of vertices, it is a complete graph.
Complete graphs have their own notation.

> The complete graph with n vertices is called K_n.

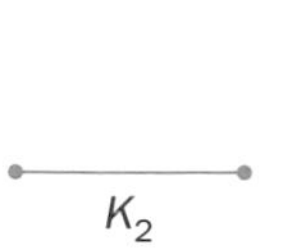

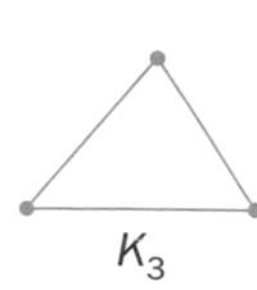

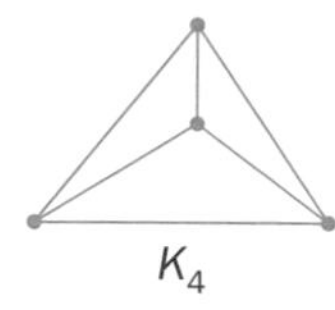

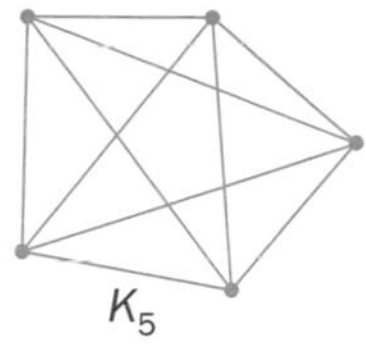

In some graphs the vertices belong to two distinct sets, and each edge joins a vertex in one set to a vertex in the other. Such a graph is called a bipartite graph.

Vertices in the same set are not directly connected.

e.g. This graph is a bipartite graph.

The two sets of vertices are $\{A, B, C\}$ and $\{D, E\}$

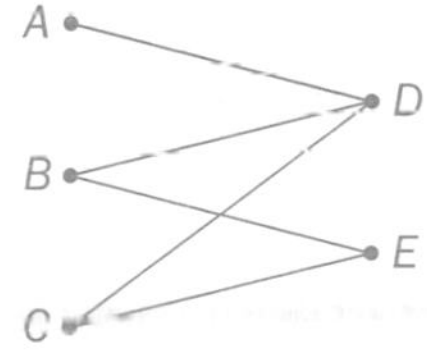

If every possible edge in a bipartite graph is present, it is a complete bipartite graph. Again there is a special notation.

> The complete bipartite graph connecting m vertices to n vertices is called $K_{m,n}$.

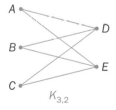

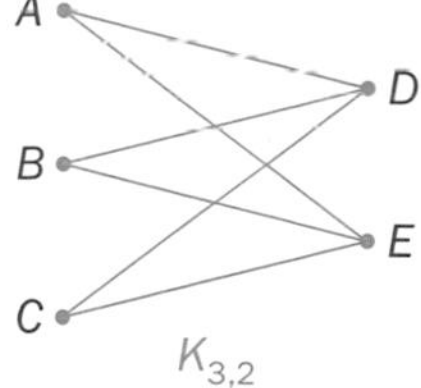

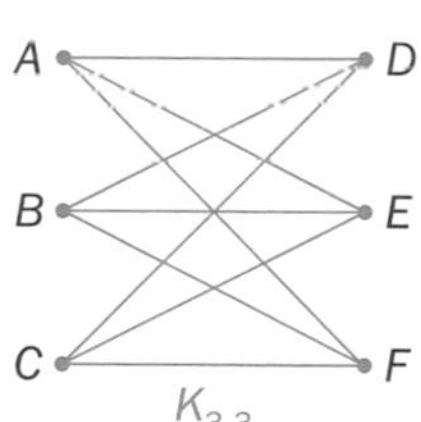

> The degree of a vertex is the number of edges which connect to it.

The terms **order** and **valency** are sometimes used in place of degree.

e.g. Each vertex in this graph has been labelled with its degree.

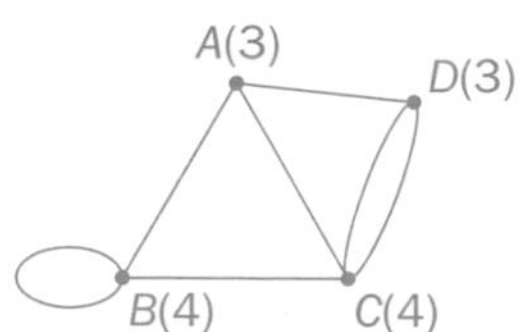

The loop at B increases its degree by 2. The degree of a vertex can be thought of as the number of 'ways out' of it.

EXAMPLE 1

Draw a graph with four vertices – one with degree 4, one with degree 2 and two with degree 1.

The solution is not unique. Here are two possible graphs.

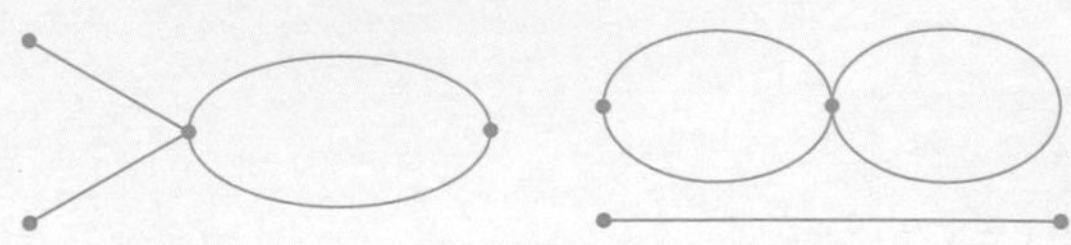

There is at least one other connected solution and one other disconnected solution. You might like to try to find these.

EXAMPLE 2

Draw a graph with vertices corresponding to the integers 2, 3, …, 9. Draw edges between vertices which have no common factor.

Write down the number of edges and the total of the degrees of the vertices.
Explain the relationship between these quantities.

Number of edges = 19

Degrees are 4, 5, 4, 7, 2, 7, 4 and 5
Total of degrees = 38

Total of degrees
= 2 × number of edges
because each edge has two ends and so contributes twice to the degree total.

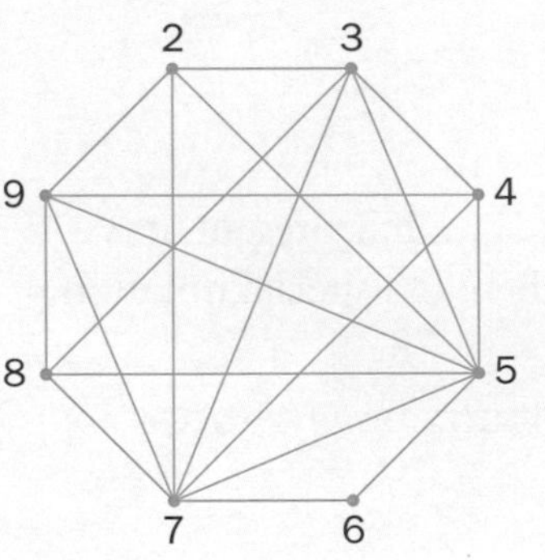

The relationship in Example 2 is true for all graphs, because each edge contributes 1 to the degree of the vertex at each of its two ends.

Total of degrees = 2 × number of edges

This is called the **handshaking lemma**, because two hands (vertices) are involved in one handshake (edge). A **lemma** is a fact which can be quoted without proof in subsequent work.

Ways of recording a graph

Instead of drawing a diagram, you can define a graph by listing the vertex set and the edge set.

e.g. The vertex set of this graph is
$\{A, B, C, D, E\}$

The edge set is
$\{AB, AC, AD, BC, BD, CD, DE\}$

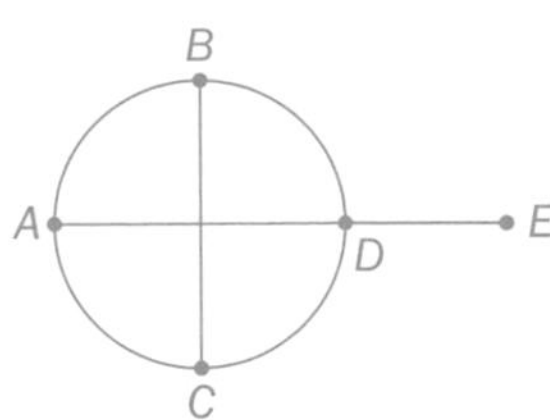

e.g. The vertex set of this graph is
{A, B, C, D}

The edge set is
{AB, AC, AD, BB, BC, CD, CD}

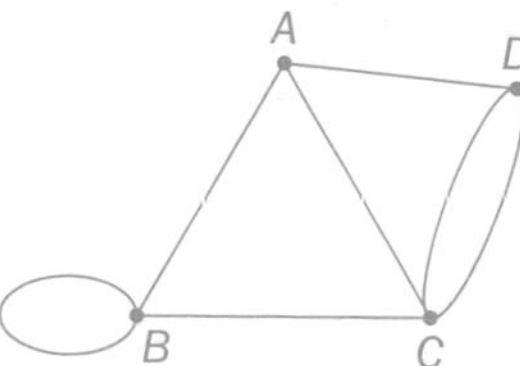

EXAMPLE 3

Draw the graph with vertex set {*A*, *B*, *C*, *D*, *E*, *F*} and edge set {*AB*, *AD*, *AE*, *BC*, *BE*, *CD*, *CE*, *DE*, *DF*, *EF*}

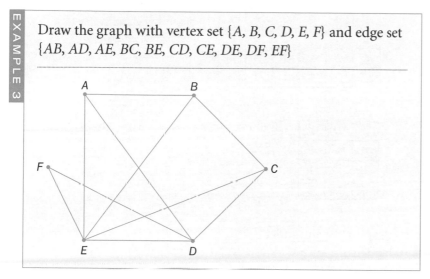

The vertex set and edge set do not tell you the layout of the diagram.
You might have drawn

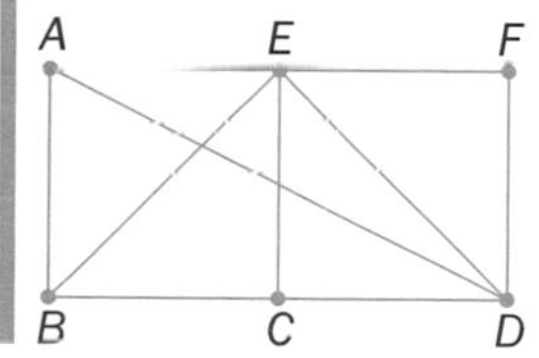

Another method of recording a graph is by means of a table – an adjacency matrix – showing the number of edges connecting the vertices.

e.g.

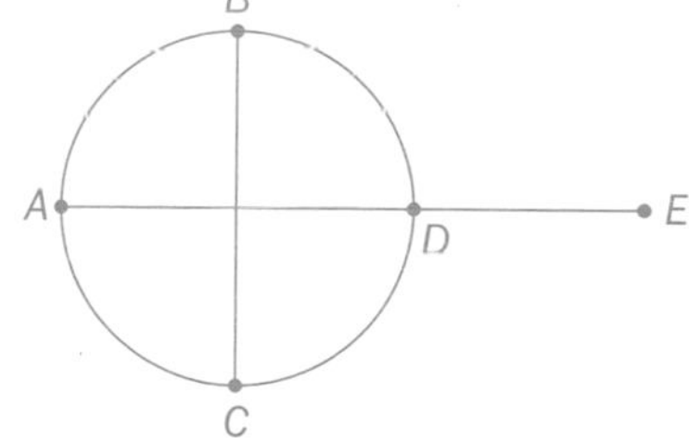

The adjacency matrix for this graph is

	A	B	C	D	E
A	0	1	1	1	0
B	1	0	1	1	0
C	1	1	0	1	0
D	1	1	1	0	1
E	0	0	0	1	0

The adjacency matrix for a graph like this is symmetrical about the leading diagonal (directed graphs are different – see Section 2.2).

e.g.

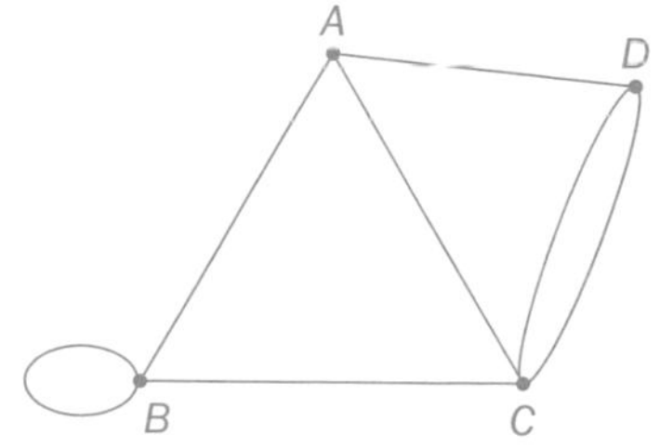

The adjacency matrix for this graph is

	A	B	C	D
A	0	1	1	1
B	1	2	1	0
C	1	1	0	2
D	1	0	2	0

The entry for *BB* is 2 because the loop could be travelled in either direction.

D1

The full adjacency matrix for a bipartite graph shows a recognisable pattern, and is often replaced by a more convenient reduced form.

e.g.

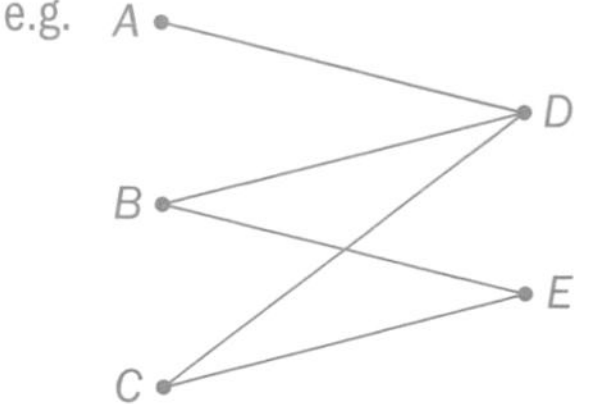

	A	B	C	D	E
A	0	0	0	1	0
B	0	0	0	1	1
C	0	0	0	1	1
D	1	1	1	0	0
E	0	1	1	0	0

Provided the vertices are listed in the right order the adjacency matrix will have two square blocks of zeros plus two symmetrical rectangular arrays.

The full adjacency matrix shown can be replaced by the reduced matrix

	D	E
A	1	0
B	1	1
C	1	1

A network, or weighted graph, is a graph with a number, called a weight, associated with each edge.
A network can be represented by means of a distance matrix, with entries corresponding to the weights of the edges.

D1

e.g. This network shows average travelling times (in minutes) between a number of towns, and the corresponding distance matrix.

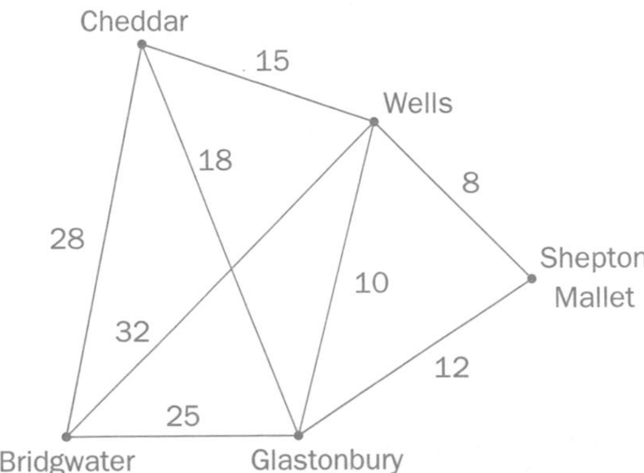

	Bridgwater	Cheddar	Glastonbury	Shepton Mallet	Wells
Bridgwater	–	28	25	–	32
Cheddar	28	–	18	–	15
Glastonbury	25	18	–	12	10
Shepton Mallet	–	–	12	–	8
Wells	32	15	10	8	–

The weights in a network may represent times, costs or other quantities, depending on the situation being modelled. Even if they are distances they may not in real life be straight lines. Because of this, the weights of a network may or may not satisfy the triangle inequality.

For any triangle, the length of each side is less than or equal to the sum of the other two sides,

so $AB \leqslant AC + BC$

However it **may not** be true that

weight $AB \leqslant$ weight $AC +$ weight BC

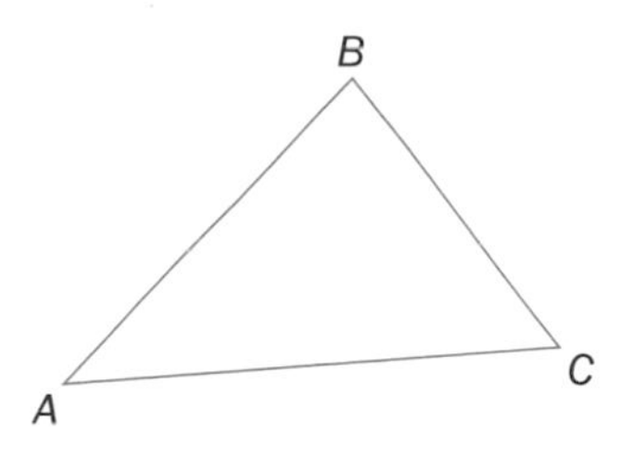

Triangle inequality

A network satisfies the triangle inequality if, for every set of three connected vertices A, B and C

weight $AB \leqslant$ weight $AC +$ weight BC

Exercise 2.1

1 Draw a connected graph which has

a one vertex of degree 3 and three vertices of degree 1

b one vertex of degree 1 and three vertices of degree 3

c two vertices of degree 3 and two vertices of degree 1

d four vertices, each with a different degree.

2 In each of these sets of four diagrams, three are isomorphic (the same graph) and the fourth is different.
Identify the odd one out.

a i  ii 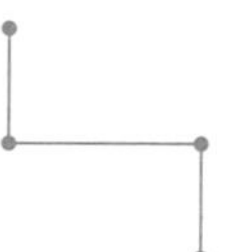 iii 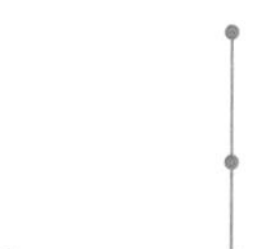 iv 

b i 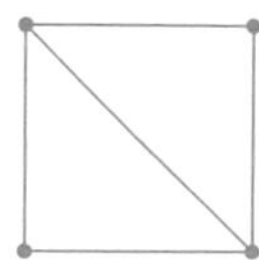 ii  iii 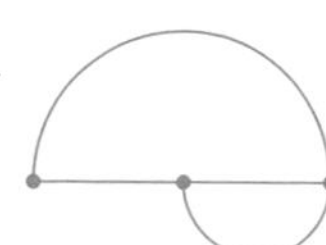 iv 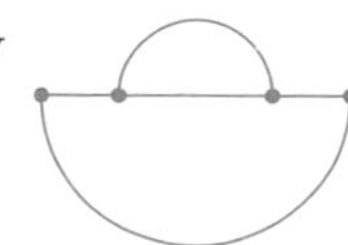

c i 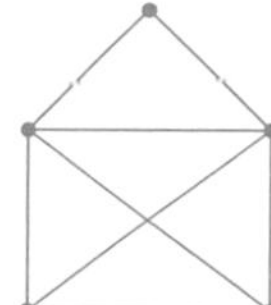 ii  iii 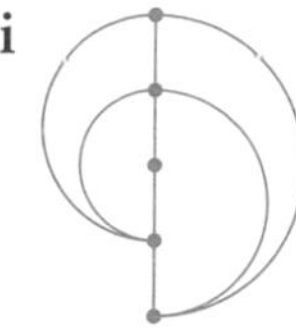 iv 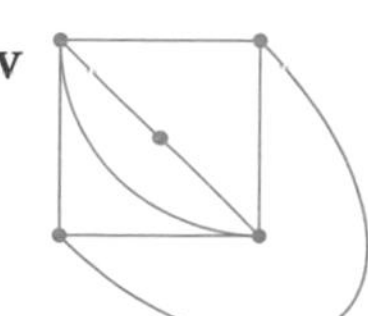

3 Draw a graph corresponding to each adjacency matrix.

a

	A	B	C	D
A	0	1	1	1
B	1	0	0	1
C	1	0	0	0
D	1	1	0	0

b

	A	B	C	D
A	0	1	2	0
B	1	0	0	1
C	2	0	0	0
D	0	1	0	2

4 Write down the adjacency matrix corresponding to this graph.

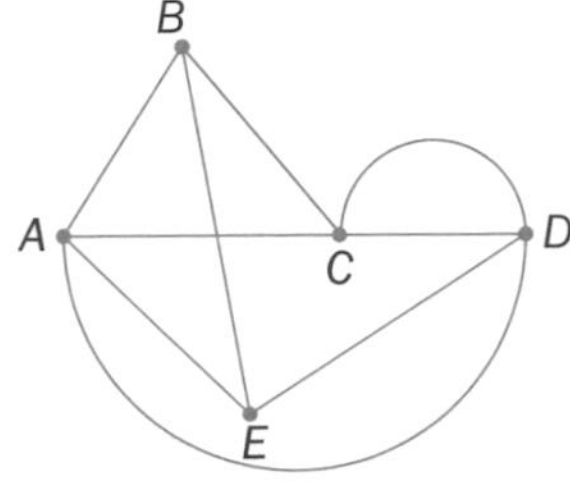

5 Write down the vertex set and edge set for each of these graphs.

a

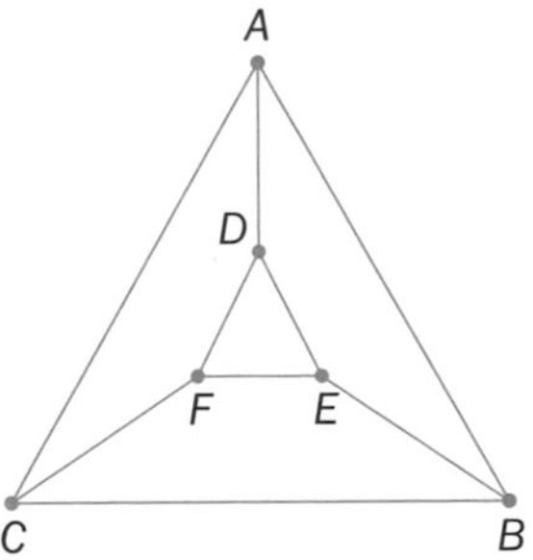

b

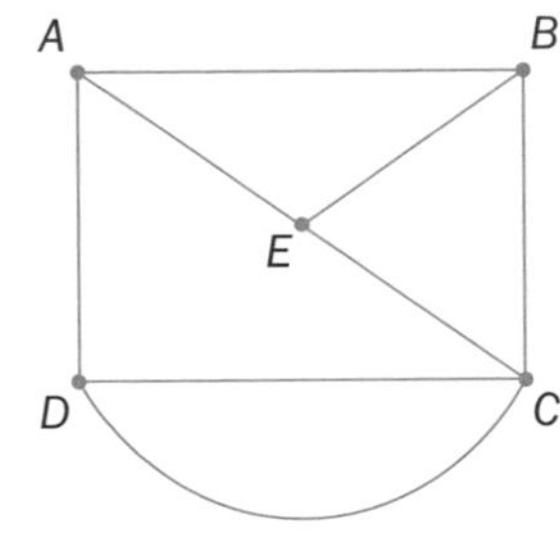

6 Draw the graph having vertex set {*A*, *B*, *C*, *D*, *E*, *F*} and edge set {*AB*, *AC*, *AE*, *BC*, *BD*, *BF*, *CE*, *DF*, *EF*, *EF*}.

7 One of these graphs is not a subgraph of the graph in question **5a**. Identify which it is, and for the others give a possible vertex set and edge set to show that it is a subgraph.

a

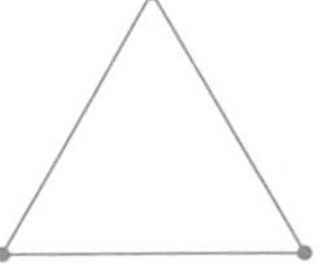

b

c

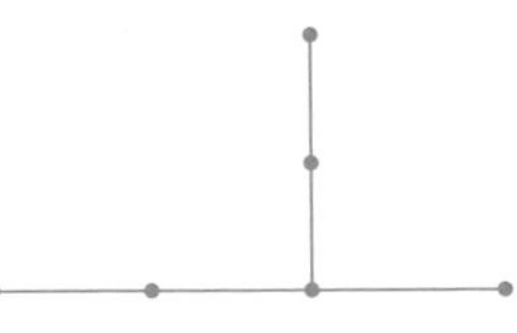

d

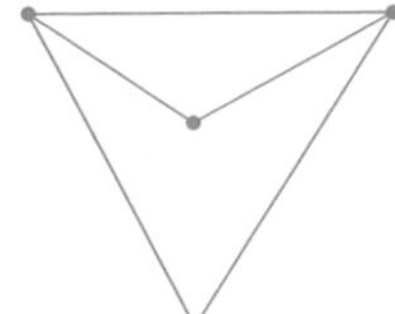

e

f

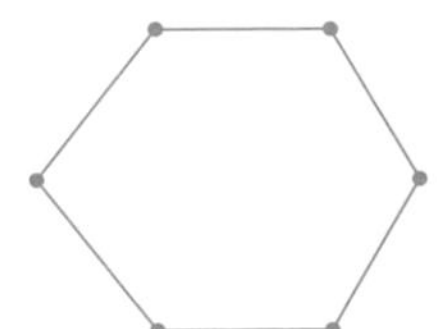

8 Draw the bipartite graph with each of these adjacency matrices.

a

	D	E	F
A	0	1	1
B	1	1	0
C	1	0	1

b

	S	T	U	V
P	1	1	0	1
Q	1	0	1	0
R	1	0	1	1

9 a Write down the adjacency matrix for the graph shown.

b By redrawing the matrix with the vertices in a different order, show that the graph is bipartite. What are the two sets of vertices involved?

c Write down the reduced version of the adjacency matrix, and redraw the graph in the more usual bipartite layout.

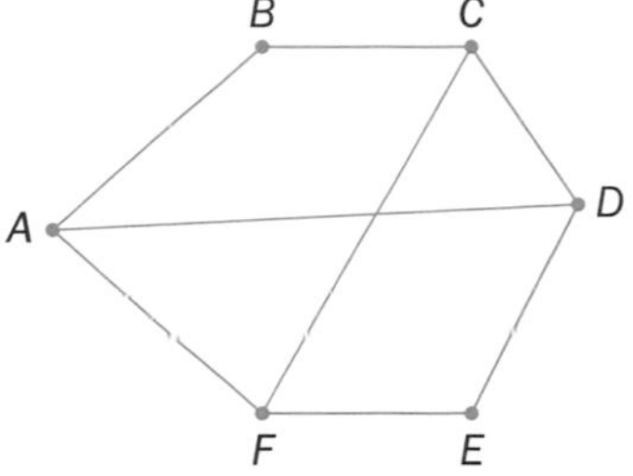

10 a Write down the degree of each vertex in the complete graph

i K_3 ii K_4 iii K_5 iv K_n

b Find the number of edges in the complete graph

i K_3 ii K_4 iii K_5 iv K_n

c Find the number of edges in the complete bipartite graph $K_{m,n}$.

d A simple bipartite graph has a total of v vertices. Find the maximum number of edges it could have.

11 a Write down the distance matrix for this network.

b Show that the network does not satisfy the triangle inequality.

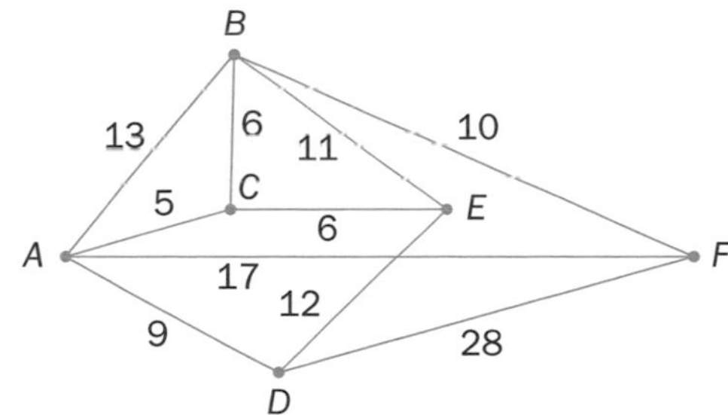

12 Draw the network corresponding to this distance matrix.

	A	B	C	D	E	F	G
A	–	60	–	26	–	55	–
B	60	–	42	64	–	–	40
C	–	42	–	–	21	–	20
D	26	64	–	–	–	35	26
E	–	–	21	–	–	75	–
F	55	–	–	35	75	–	70
G	–	40	20	26	–	70	–

2.2 Directed graphs and networks

Sometimes specific edges of a graph are allocated a direction of travel. These edges are called **directed edges**. The graph is a **directed graph** or **digraph**.

Directed edges are like one-way streets in real life.

A **digraph** (**directed graph**) has at least one edge with an associated direction.

e.g. The graph might represent a network of pipes in which the liquid flows in one direction only.

You can represent a digraph using an adjacency matrix, but it must use 'from' and 'to' to show the direction of the edges.

EXAMPLE 1

Write down the adjacency matrix for the digraph shown.

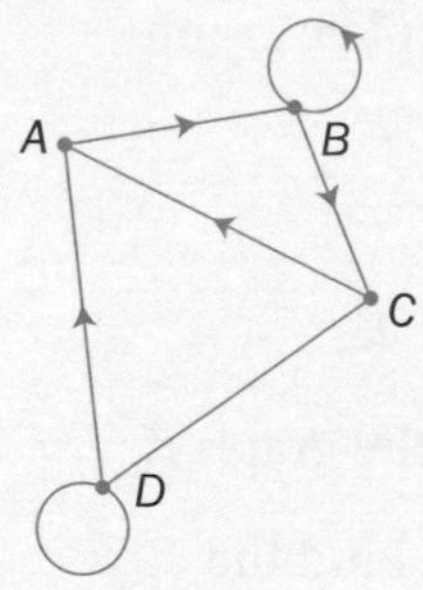

The adjacency matrix for a digraph will not be symmetrical, unlike that for an undirected graph.

		To			
		A	*B*	*C*	*D*
From	*A*	0	1	0	0
	B	0	1	1	0
	C	1	0	0	1
	D	1	0	1	2

The entry for *BB* is 1 because the loop is directed, but for *DD* the entry is 2 because the loop can be travelled in either direction.

You can also represent a digraph by stating its vertex set and edge set.

EXAMPLE 2

Write down the vertex set and edge set for the digraph in Example 1.

Vertex set {*A*, *B*, *C*, *D*}
Edge set {*AB*, *BB*, *BC*, *CA*, *CD*, *DA*, *DC*, *DD*, *DD*}

The order of letters in the edge set is now important. *CD* and *DC* are both listed because the edge is **bidirectional**. Similarly *DD* is listed twice.

A **directed network** is a weighted graph with directed edges. The corresponding distance matrices will be asymmetrical.

Exercise 2.2

1 For each digraph write down

i the vertex set and edge set

ii the adjacency matrix.

a

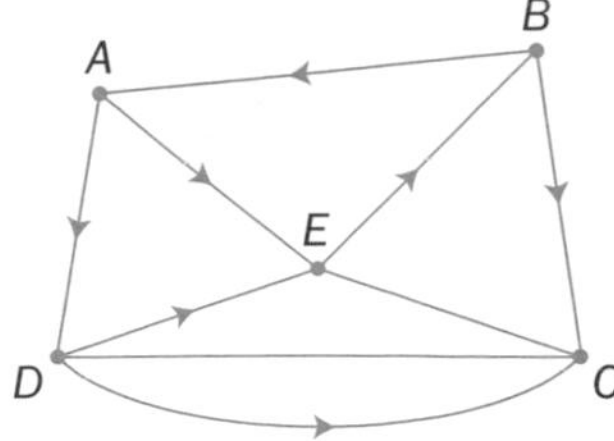

b

P
Q
T
S
R

2 Draw the digraph corresponding to this adjacency matrix.

		To			
		A	B	C	D
From	A	0	1	2	0
	B	0	1	1	1
	C	1	0	0	1
	D	1	0	1	0

3 Draw the digraph corresponding to the vertex set {V, W, X, Y, Z} and the edge set {VW, VY, VZ, VZ, WX, WY, XX, XY, YV, YZ, ZX}.

4 Write down the distance matrix for this network.

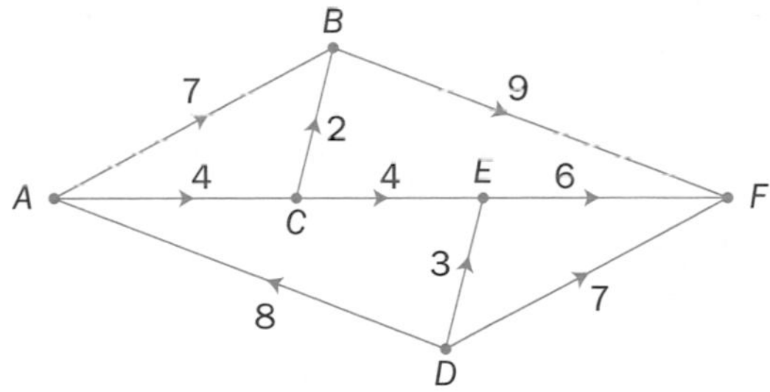

5 Draw the directed network corresponding to the distance matrix given.

		To				
		A	B	C	D	E
From	A	–	8	5	–	9
	B	8	–	–	–	–
	C	–	3	–	6	–
	D	2	4	11	–	–
	E	–	–	5	–	–

2.3 Modelling

The real world is complicated. Exploring a situation mathematically is only possible if you simplify things. This means making assumptions about which factors to include in your calculations and which factors you can ignore.

This simplified setup is a **mathematical model** of the situation. The predictions from this model can be compared with the real world outcomes, and if necessary the assumptions can be modified to give a better model.

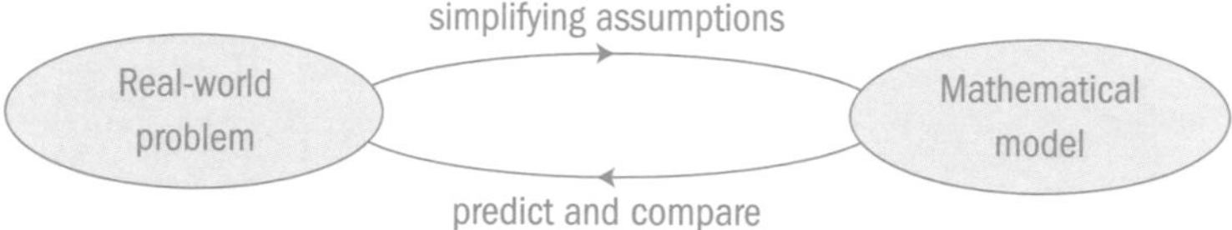

Graphs and networks give a simplified representation of a wide variety of situations.

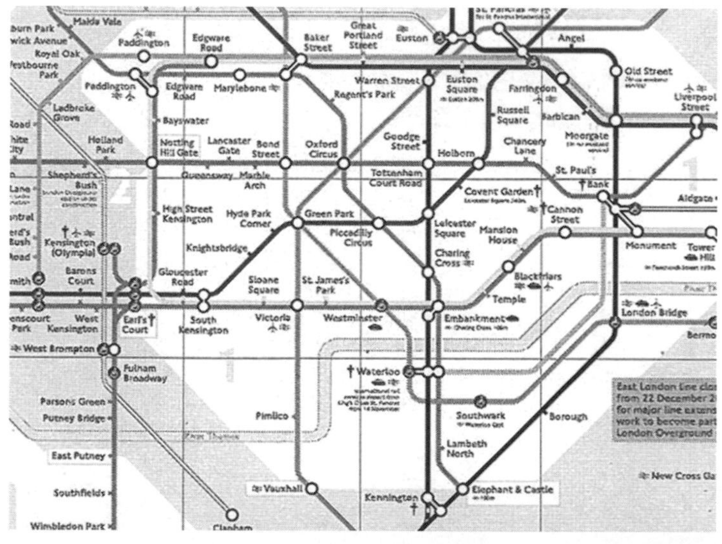

e.g. The map of the London Underground is a graph used as a model. The actual geography of the system and the distances between stations are not accurately represented, because the user only needs to know which line to travel and which station follows which.

EXAMPLE 1

A sociological study of friendships finds that Anna is friends with Benji, Chloe and Dai. Benji is friends with Dai, but neither of them knows Chloe. Represent this as a graph.

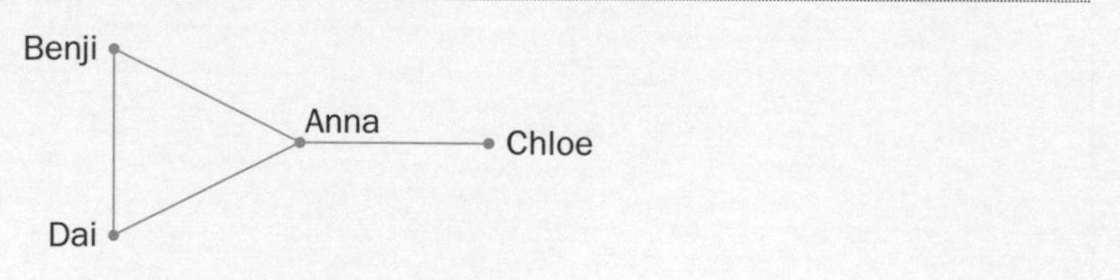

D1

Example 2 requires a digraph.

EXAMPLE 2

A carpenter making a trolley lists the tasks involved. These are

A – cut out the base
B – cut out the sides
C – attach castors
D – drill screw holes in the base
E – fix the sides together
F – fix the sides to the base
G – paint the trolley

Task A must be completed before tasks C and D can take place. Task B must precede task E. F cannot happen until D and E are done. G depends on the completion of C and F. Show this as a graph.

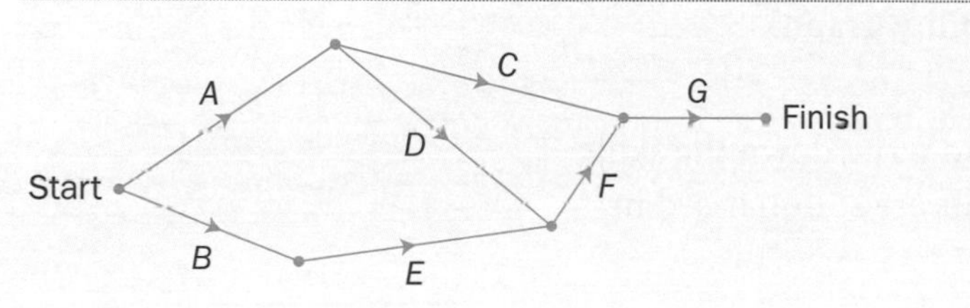

Problems like this are covered in detail in Chapter 8.

If the duration of each task were known the times could be included as weights, making this a directed network.

Example 3 makes use of a bipartite graph.

EXAMPLE 3

Ice cream is available in four flavours – strawberry, vanilla, chocolate and mint. George likes strawberry and chocolate, Harriet likes vanilla, chocolate and mint, Isla likes strawberry and mint while Juan only likes chocolate.
Show this information as a graph.

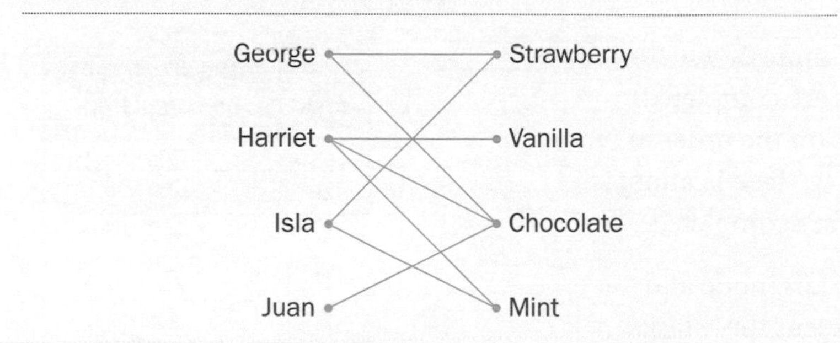

Situations like this are covered in detail in Chapter 6.

Example 4 shows how an architect might use a graph as a model.

EXAMPLE 4

The plan shows a building with five rooms and two doors to the outside. Draw a graph to show how a person can move from room to room and from the building to the outside.

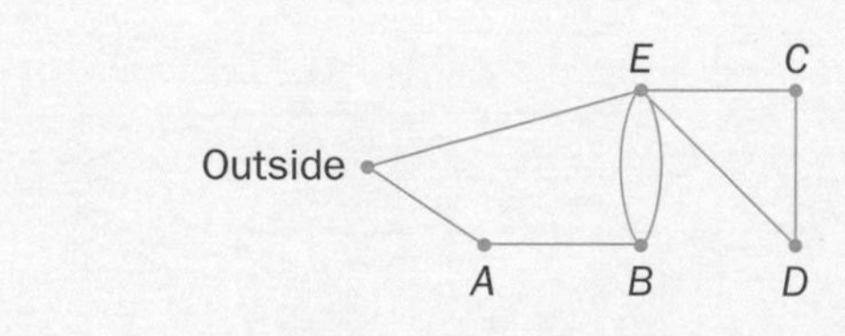

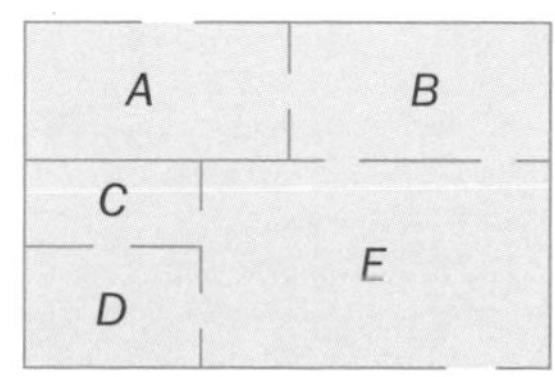

This is known as a **circulation graph**.

Exercise 2.3

1 A team of four runners is to run a 4 × 100 m relay race. Dwayne likes to run either the first or second leg. Anton will run any leg other than the first. Kris prefers the third or fourth leg. Jason likes either the first or the last leg. Draw a bipartite graph to show this information.

2 The diagram shows a vegetable patch with seven beds. The owner wants to plan the planting so that beds next to each other do not contain the same type of plant.

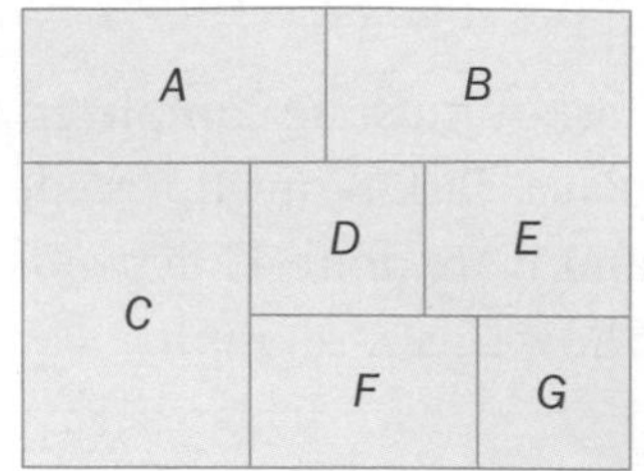

Draw a graph with a node corresponding to each bed. Draw arcs between those beds which could contain the same type of plant (this is known as a compatibility graph).

3 A group of five employees are asked to write down the skills in which they are qualified. The results are summarised in the table.

Name	Skills
Brown	Bricklaying, carpentry, painting
Choli	Electrical installation, plumbing
Derry	Plumbing, glazing, painting
Everest	Bricklaying, electrical installation, glazing
Farr	Plumbing, electrical installation, painting

Draw a bipartite graph to show this information.

4 A project involves eleven tasks, some of which cannot be started before others are completed. The table shows the limitations on the order in which the tasks can be done – e.g. Task I cannot be started before tasks E and F are completed.

Draw a directed graph, with a 'start' node and a 'finish' node, whose arcs represent these tasks.

Task	Tasks which must be completed
A	–
B	A
C	A
D	B
E	B
F	C
G	C
H	D
I	E, F
J	G
K	H, I

D1

5 The diagram shows the layout of a building with seven rooms. There are three doors to the outside.

Draw a circulation graph, with vertices for the rooms and the outside, and edges showing possible movement between them.

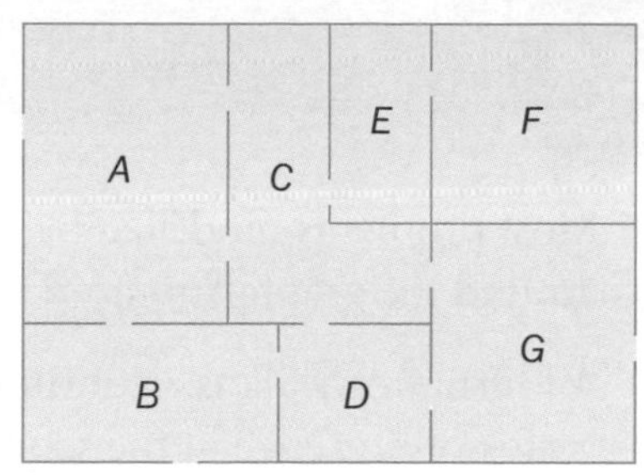

6 Ecologists use digraphs (called a food web) to show which species feed on which. Here is an example for some plants and animals in a garden.

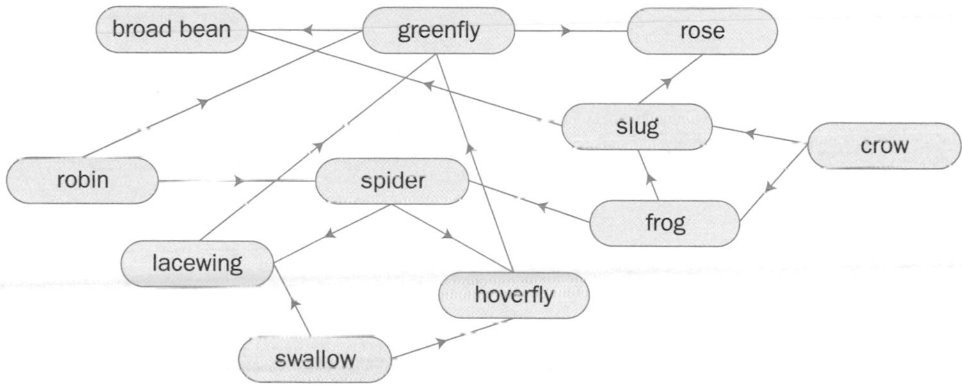

From this graph, identify

a which species feed on broad beans

b which species are eaten by frogs

c which species are in competition with i robin ii spider.

7 The diagram shows a road junction. Traffic lights are to be installed, with a sequence designed to allow as many streams of traffic as possible to flow at any given time.

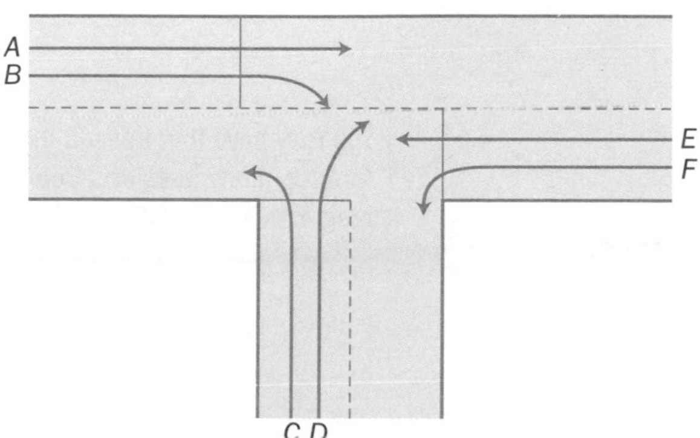

Draw a graph with a vertex for each traffic stream to show which streams can safely flow at the same time (this is another example of a compatibility graph).

2.4 Moving around a graph

Most problems modelled by graphs and networks involve moving around the graph. You need to know the terminology used.

A journey or walk is a continuous sequence of edges. The end vertex of one edge in the sequence is the start vertex of the next.

> A path is a walk which does not pass through any vertex more than once.

If vertices are visited more than once the journey is called a **trail**.

e.g. *A B C F E* is a path.

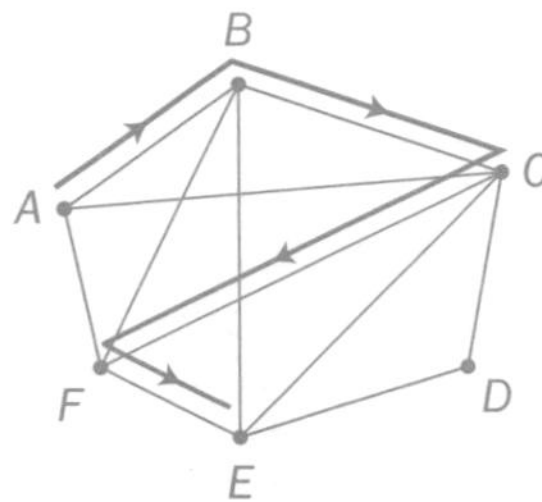

A journey which returns to its start vertex is said to be closed.

> A closed path is called a cycle or circuit.

Decision Mathematics is quite a recent branch of the subject, and you may find other definitions of these terms. For the present syllabus you need to know the terms **path** and **cycle** (**circuit**) as defined here.

e.g. *A B E F C A* is a cycle.

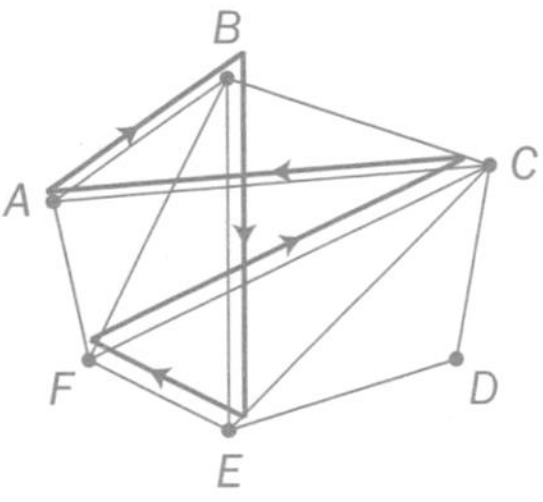

Some graphs have no cycles.

> A connected graph with no cycles is a tree.

e.g. This graph is a tree.

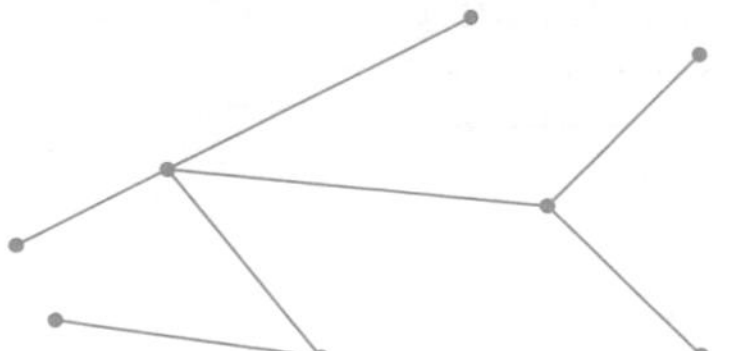

You may have met trees in the form of family trees and tree diagrams for probability.

D1

Exercise 2.4

1 The diagram shows a graph.

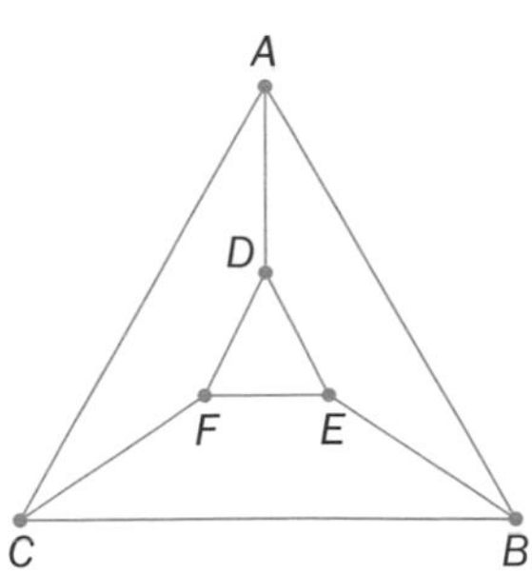

a Write down a path from A to C which uses

i 2 edges ii 3 edges
iii 4 edges iv 5 edges.

b Write down a cycle, starting and ending at B, which involves (including B)

i 3 vertices ii 4 vertices
iii 5 vertices iv all 6 vertices.

c Explain why $EFCABC$ is not a path.

2 Which of these graphs are trees?

a 

b 

c 

d 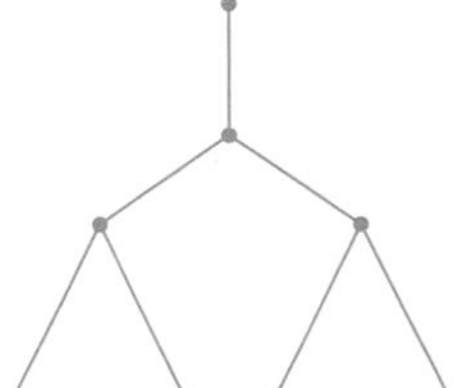

e 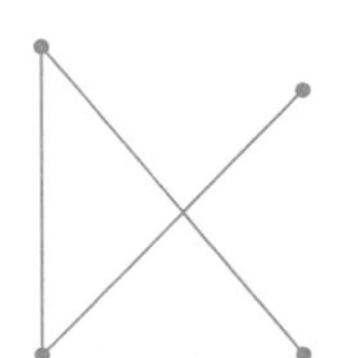

f 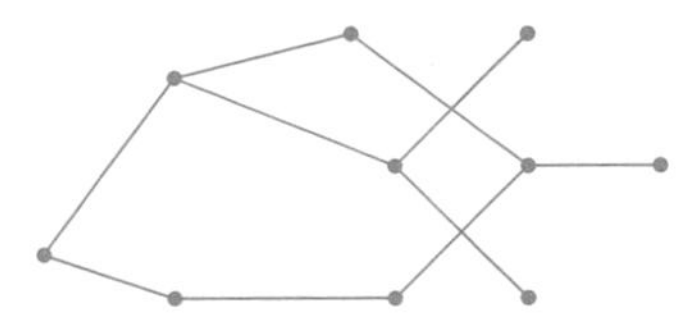

3 There are only two distinct trees with 4 vertices.

These are and

All other such trees are isomorphic to one of these two.

How many distinct trees can you draw with

a 5 vertices b 6 vertices?

Traversing a graph

You may have seen puzzles in which you try to draw a graph without lifting your pencil or going over a line twice.

e.g. You can draw this graph as shown.

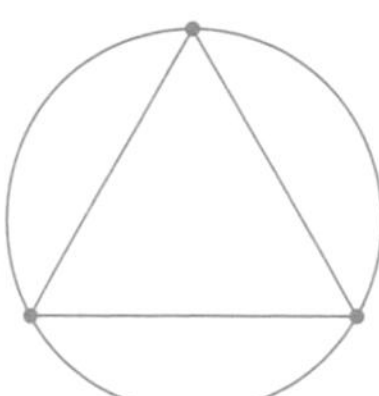

The solution given starts and finishes at the same place. This is true of every solution for this graph.

You might like to try to find your own solution.

Graphs like the one in this example are called traversable or Eulerian (pronounced *oi-leerie-ann*) graphs.

Leonhard Euler (1707–1783) was a Swiss mathematician. The study of traversable graphs was one of his many important contributions to mathematics.

A graph is traversable or Eulerian if it has a trail which is closed (starts and ends at the same vertex) and which includes every edge once and once only.

The trail is called an **Eulerian trail**.

Some graphs cannot be traversed.

e.g. Try to draw this graph. You should find that it is not possible unless you go over a line twice or take your pencil off the page.

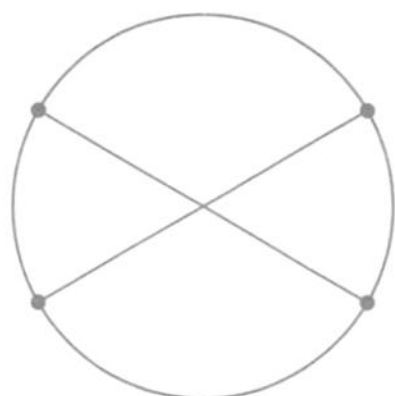

There are some graphs that you can only draw by starting and ending at different points. Graphs like this are semi-traversable or semi-Eulerian.

e.g. You can draw this graph as shown.

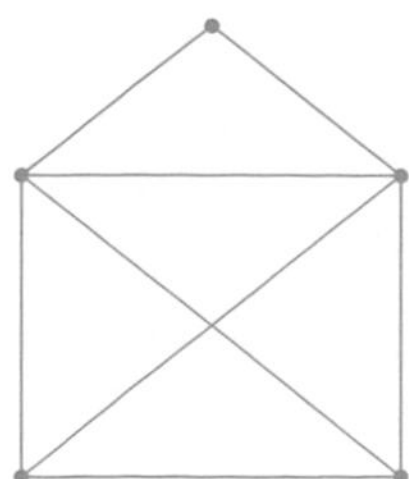

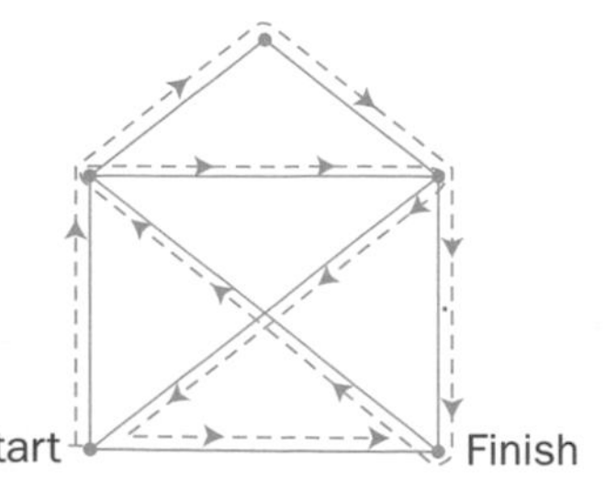

Try your own solution to this. You should find that it is only possible if you use the two bottom vertices as the start and finish points.

You can tell if a graph is traversable by examining the degree of its vertices. What matters is whether vertices have **odd degree** or **even degree**.

The degree of a vertex is the number of edges which connect to it (see Section 2.1).

A graph is **traversable (Eulerian)** if all its vertices are of **even degree**.

e.g. *ABC* is Eulerian because all its vertices are even (degree 4).

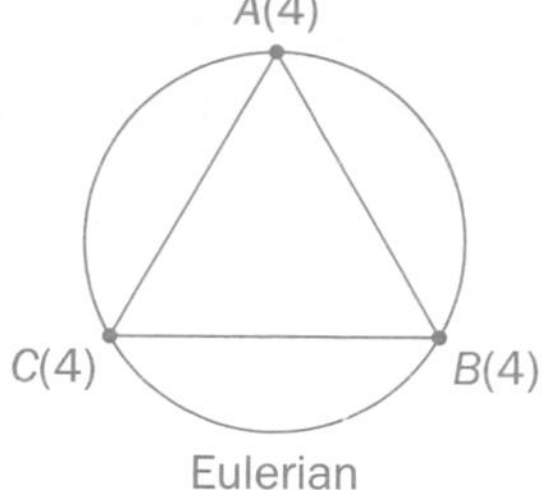

A graph is **semi-traversable (semi-Eulerian)** if it has two odd vertices.
The two odd vertices are the start and finish vertices.

e.g. *DEFGH* is semi-Eulerian, with *D* and *E* as the start and finish vertices.

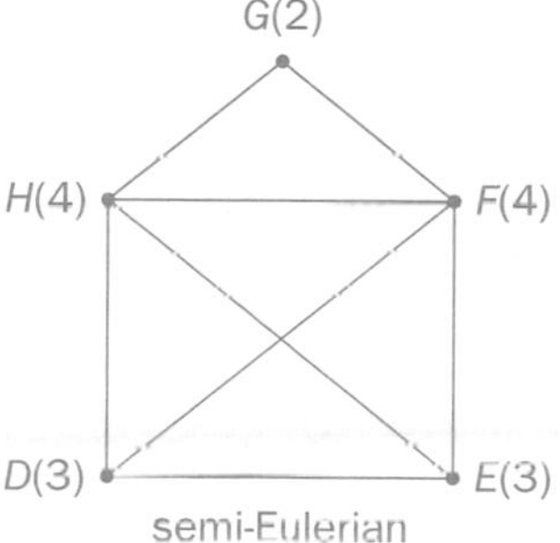

A graph with more than two odd vertices is not traversable.

e.g. *IJKL* is not traversable because all four of its vertices are odd.

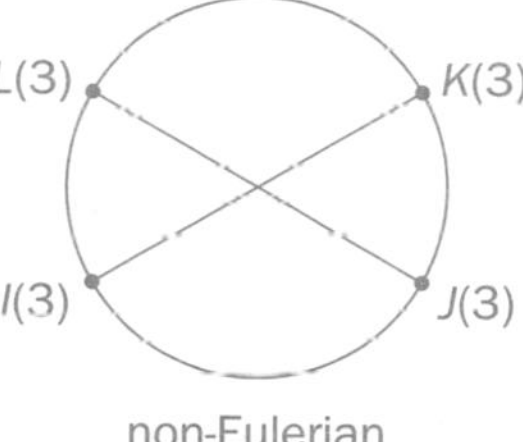

The number of odd vertices in a graph is even.

This follows because

total of degrees = 2 × number of edges

so the total of the degrees must be an even number.

This is the **handshaking lemma** (see Section 2.1).

To get an even total you must have an even number of odd values in the list.

Proof of the conditions for a graph to be Eulerian

When an Eulerian or semi-Eulerian trail goes through a vertex, there is an 'incoming' edge and an 'outgoing' edge. This means that, apart from the start/finish points, every vertex must be an even vertex.

If the graph is Eulerian, the start and finish are at the same vertex. This will have an even number of edges plus the starting edge and the finishing edge, so it will be an even vertex.

If the graph is semi-Eulerian, start and finish vertices are different. They must each have an even number of edges plus either the starting or finishing edge, so they must both be odd vertices.

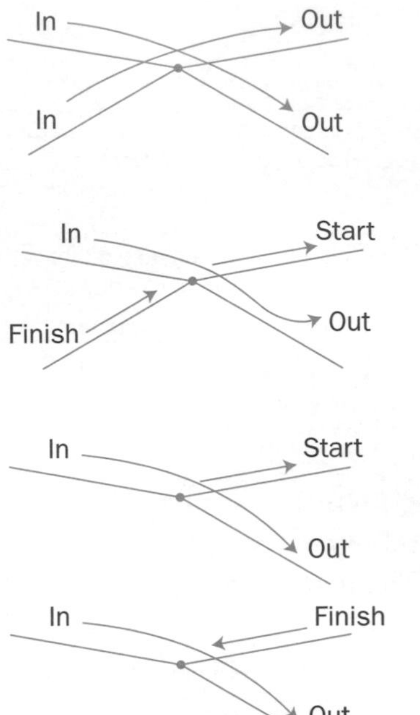

EXAMPLE 1

Explain why the graph shown is Eulerian and give an example of an Eulerian trail for this graph.

The degrees of the vertices are $A(4)$, $B(4)$, $C(4)$, $D(4)$, $E(4)$ and $F(2)$.

All the vertices have even degree, so the graph is Eulerian.

A possible Eulerian trial is *ABEFDABCEDCA*.

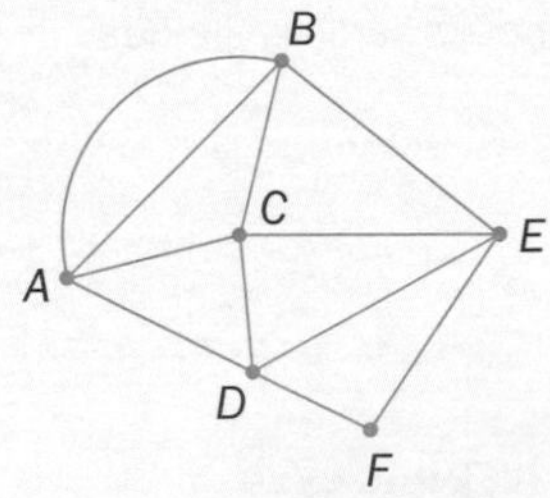

There is an algorithm, called Fleury's algorithm, for finding an Eulerian trail, but it is outside the present syllabus.

EXAMPLE 2

Determine whether the graph shown is Eulerian, semi-Eulerian or neither. If possible, find a trail which traverses the graph.

The degrees of the vertices are $A(3)$, $B(4)$, $C(5)$, $D(4)$, $E(4)$ and $F(4)$.

There are two odd vertices, so the graph is semi-Eulerian. The start and finish vertices are A and C.

A possible trail is *ABEFDACEFCBDC*.

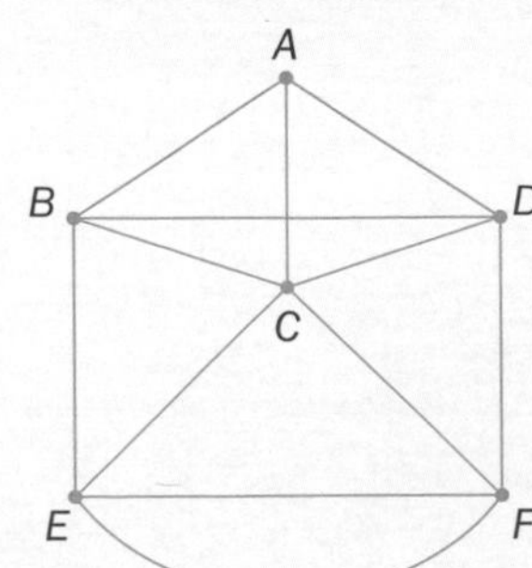

Exercise 2.5

1 State whether each of these graphs is Eulerian, semi-Eulerian or neither.

a
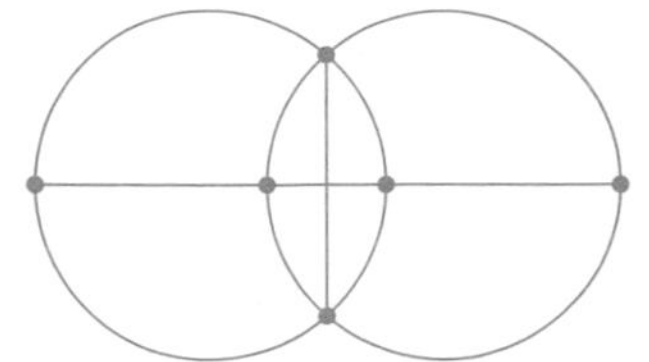

b
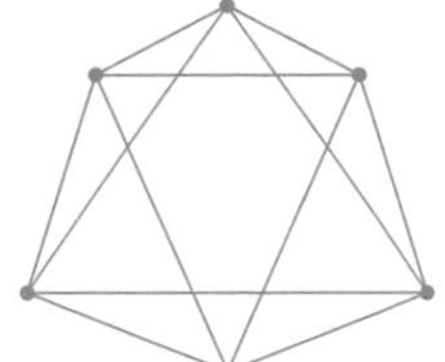

c
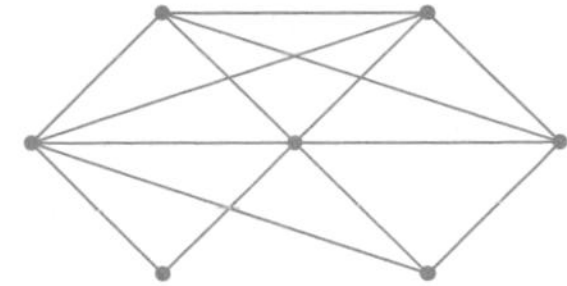

d
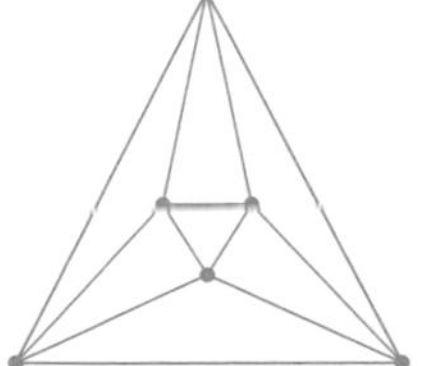

e
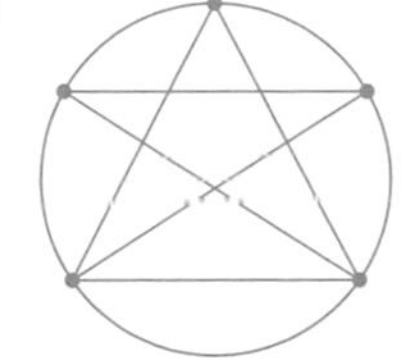

f
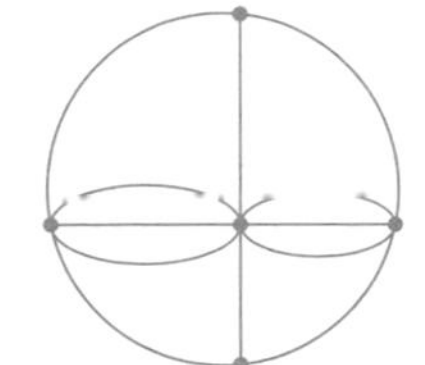

2 a Draw the graph corresponding to the adjacency matrix shown.

b Explain why you know that the graph is Eulerian.

c State a possible Eulerian trail for the graph.

d Explain how you could tell that the graph is Eulerian by referring to the matrix and without drawing the graph.

	A	*B*	*C*	*D*	*E*	*F*	*G*
A	–	1	1	0	0	0	0
B	1	–	1	1	1	0	0
C	1	1	–	1	0	1	0
D	0	1	1	–	1	1	0
E	0	1	0	1	–	1	1
F	0	0	1	1	1	–	1
G	0	0	0	0	1	1	–

3 By examining the complete graphs K_3, K_4, K_5 and K_6 or otherwise, state the conditions under which the complete graph K_n is traversable.

4 The diagram shows the town of Königsberg. There is an island just downstream from where two rivers meet, and in the eighteenth century there were seven bridges, as shown.

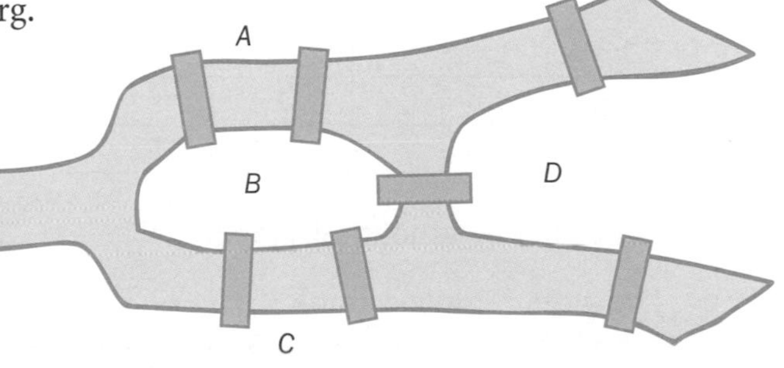

The citizens of Königsberg used to spend their Sunday afternoons trying to walk across each bridge once only, ending up at their starting point.

a Representing the land regions *A*, *B*, *C* and *D* as vertices and the bridges as edges, draw a network to model the situation.

b State whether the citizens of Königsberg were wasting their time trying to solve this problem.

This problem, the Königsberg bridge problem, was part of Euler's original study of the problem of traversability, published in 1736

Review 2

1 For the graph shown in the diagram

a state the degree of each vertex

b state the number of edges and explain the relationship between this and the sum of the values in part **a**

c write down the vertex set and the edge set

d write down the adjacency matrix. To what do the sums of the columns in the adjacency matrix correspond?

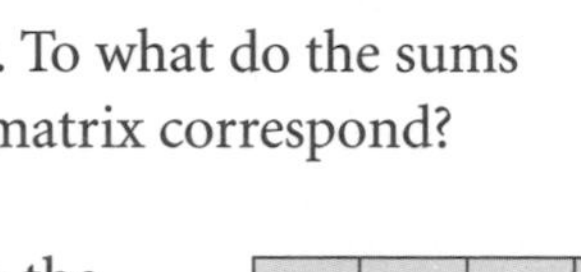

2 a Draw the graph corresponding to the adjacency matrix shown.

b Is the graph a connected graph? Explain your answer.

c Is the graph a simple graph? Explain your answer.

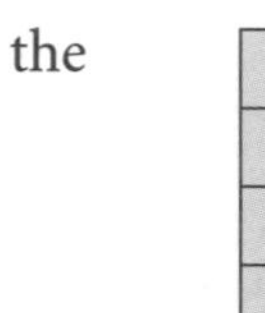

	A	B	C	D	E	F
A	0	1	0	0	1	0
B	1	0	0	0	2	0
C	0	0	0	1	0	1
D	0	0	1	0	0	1
E	1	2	0	0	0	0
F	0	0	1	1	0	0

3 Alan has four friends – Ben, Candy, Dee and Eli. Ben is friends with Candy and Eli but doesn't know Dee. Candy is friends with Dee and Eli. Dee does not know Eli.

a Copy and complete the adjacency matrix showing these friendships.

	Alan	Ben	Candy	Dee	Eli
Alan	0	1	1	1	1
Ben	1	0	1	0	1
Candy	1				
Dee	1				
Eli	1				

b Draw the corresponding graph.

c Draw the complete graph K_4.

d Is the graph K_4 a subgraph of the graph you drew in part **b**? If it is, list the vertices corresponding to this subgraph.

4 From the graphs shown, identify

a those which are not simple graphs

b those which are not connected graphs

c graphs which are isomorphic to each other.

i

ii 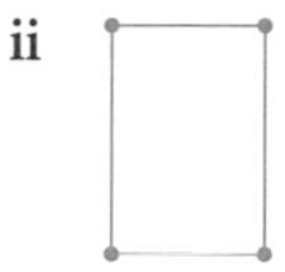

iii 

iv 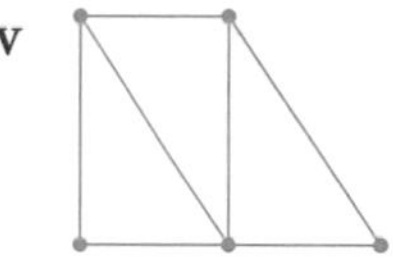

v 

vi 

vii 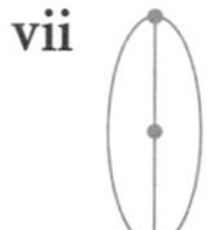

viii 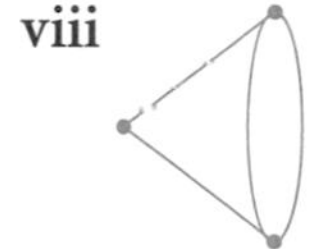

ix 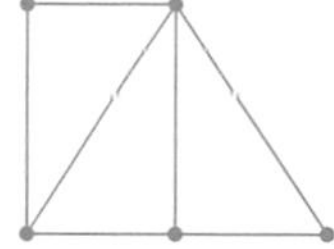

x 

5 This adjacency matrix corresponds to a bipartite graph.

a Draw the graph.

b Draw the complete bipartite graph $K_{3,3}$.

c Is the graph $K_{3,3}$ a subgraph of the graph you drew in part a? If it is, list the vertices involved.

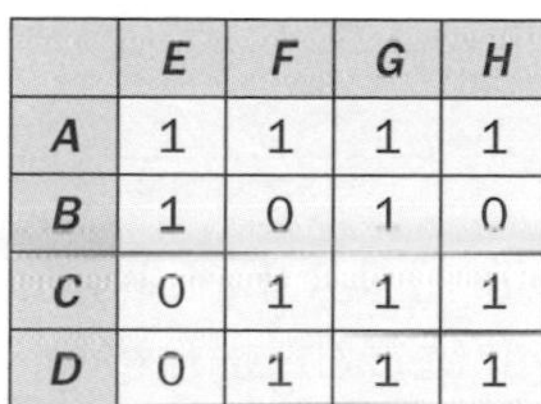

	E	*F*	*G*	*H*
A	1	1	1	1
B	1	0	1	0
C	0	1	1	1
D	0	1	1	1

6 Four friends – John, Kim, Lee and Mel – wish to form a band using guitar, bass, keyboard and drums. The table lists the skills they have.

a Draw a bipartite graph to show this information.

b Can they form the band without someone learning new skills? Explain your answer.

Name	Can play
John	Guitar, keyboard
Kim	Guitar, keyboard
Lee	Bass, keyboard, drums
Mel	Keyboard

7 The diagram shows the direct bus routes offered by a coach operator, together with the cost, in pounds, of a ticket on each route.

a Draw a distance matrix corresponding to this network.

b Does the network satisfy the triangle inequality? Explain your answer.

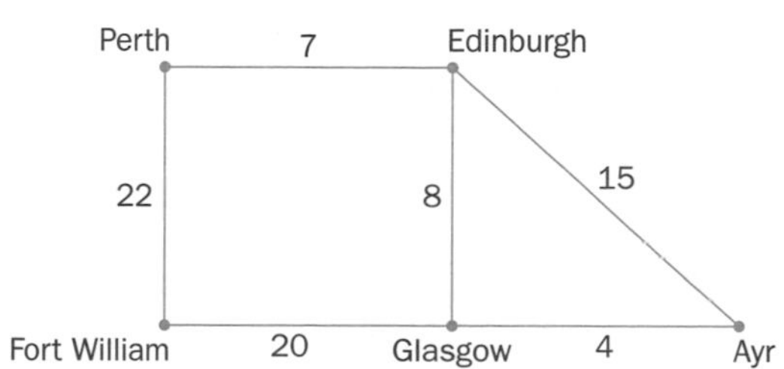

8 For the digraph shown

a list the vertex set and the edge set

b construct an adjacency matrix.

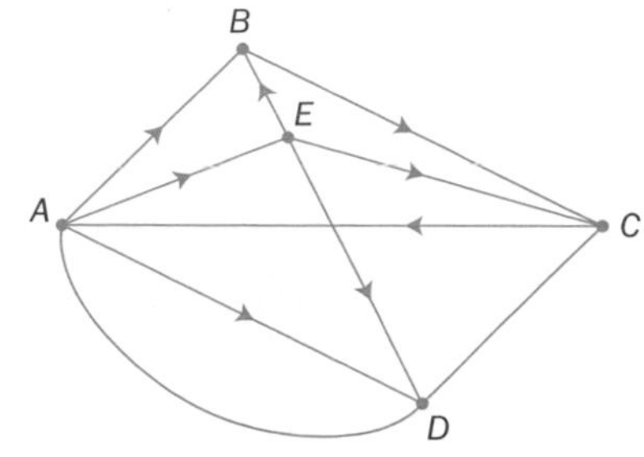

9 Draw the digraph corresponding to this adjacency matrix.

		To				
		A	*B*	*C*	*D*	*E*
From	*A*	0	1	0	0	2
	B	0	1	0	1	1
	C	1	1	0	1	0
	D	0	0	2	0	0
	E	1	0	1	1	2

10 Construct a distance matrix for the directed network shown.

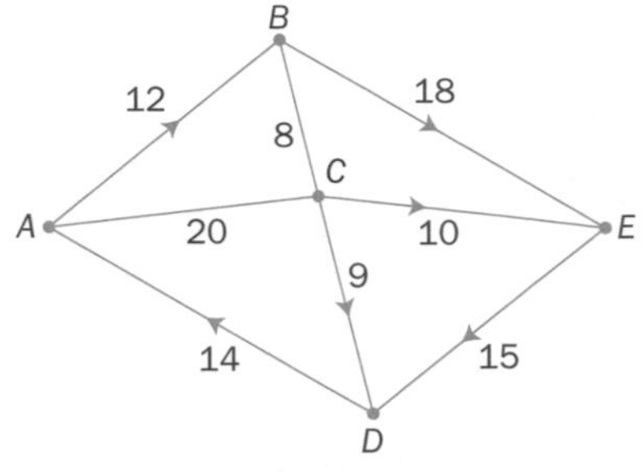

11 Anneka makes the following statements about the graph shown. In each case, state, with reasons, whether her statement is true or false.

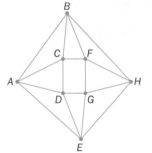

a *ABCDGFCB* is a path.

b *HFCAEH* is a cycle.

c The graph is not traversable.

12 The structure of a saturated hydrocarbon molecule is a tree with nodes of degree 4 (the carbon atoms) and nodes of degree 1 (the hydrogen atoms). Draw two distinct molecules with four carbon atoms. How many hydrogen atoms are required in each case?

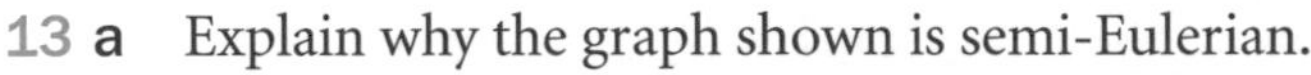

13 a Explain why the graph shown is semi-Eulerian.

b Nathan adds one edge to the graph, and then claims it is Eulerian. Assuming he is right, which edge did he add?

c Give an example of an Eulerian trail for Nathan's graph.

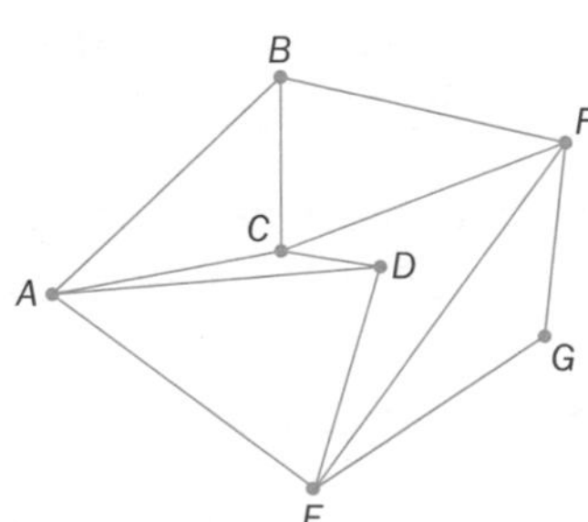

14 The diagram shows a building with five rooms. There is a door through every wall between rooms and between the rooms and the outside.

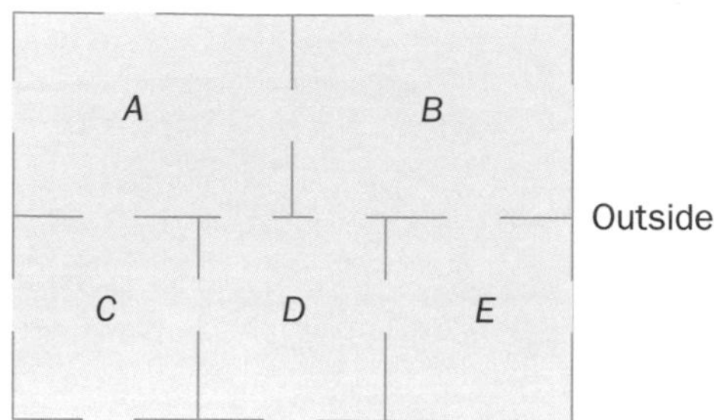

a By representing the rooms and the outside as vertices and the doors as edges, illustrate this situation as a graph.

b Demelza tries to find a route which passes through every door and finishes at its starting point.
Explain, with reference to your graph, why such a route is not possible.
George locks some of the doors so that Demelza can find her desired route through the remaining doors.

i What is the smallest number of doors he could lock?
ii Suggest a possible minimum set of doors.

15 A simple graph, *G*, has six vertices.

a If *G* is a tree, how many edges does it have?

b If *G* is a complete graph, how may edges does it have?

c If *G* is a bipartite graph, what is the maximum number of edges it could have?

d What is the maximum number of edges there could be in the longest cycle in *G*?

e If *G* has a cycle with three edges, explain why *G* cannot be bipartite.

16 a Draw the graph corresponding to the adjacency matrix shown.

	A	*B*	*C*	*D*	*E*
A	0	1	1	0	1
B	1	0	1	0	1
C	1	1	0	1	1
D	0	0	1	0	1
E	1	1	1	1	0

b i List all possible paths from *A* to *D*.
ii List all possible circuits from *A* which pass through every vertex of the graph.

2 Exit

Summary

Refer to

- A graph is a set of points (vertices or nodes) connected by lines (edges or arcs). You need to know the vocabulary of graphs. 2.1
- You can describe a graph by listing its vertex set and edge set, or by using an adjacency matrix. 2.1
- A network is a graph whose edges have associated values, called weights. You can describe a network by using a distance matrix. 2.1
- A directed graph (digraph) has some or all of its edges with an associated direction. Similarly, a network with directed edges is a directed network. 2.2
- You can use graphs and networks to model a variety of problems. 2.3
- You need to know the vocabulary associated with moving around a graph. 2.4
- A graph is traversable (Eulerian) if there is a closed trail passing along every edge once only. Every vertex of an Eulerian graph has an even degree. 2.5

Links

Graphs are widely used in analysing the relationships within ecosystems. The different elements within the system, such as cultivated crops, other plants, pests, predators, soil conditions and weather can be represented as vertices of a graph. A directed edge joins two elements if the first has an effect on the second, with a weight indicating the strength of the effect. Such a directed network helps in the study of the likely chain of consequences resulting from a change in one of the elements.

3 Minimum spanning trees

This chapter will show you how to
- understand the terms spanning tree and minimum spanning tree (minimum connector)
- identify the minimum spanning tree from a network diagram using Kruskal's algorithm and Prim's algorithm
- identify the minimum spanning tree from a distance matrix using Prim's algorithm.

Introduction

If you were manager of a nature reserve with a number of wildlife observation hides in an area of marshland, you would need to construct walkways between the hides. Your aim would be to build the least amount of walkway that would let people travel between any two hides.

You will learn how to solve problems of this type in this chapter.

3.1 Spanning trees

Many of the subgraphs of a given graph are trees. Some of these, called spanning trees, contain every vertex of the graph.

A spanning tree or connector of a graph is a subgraph which is a tree containing every vertex of the graph.

e.g.

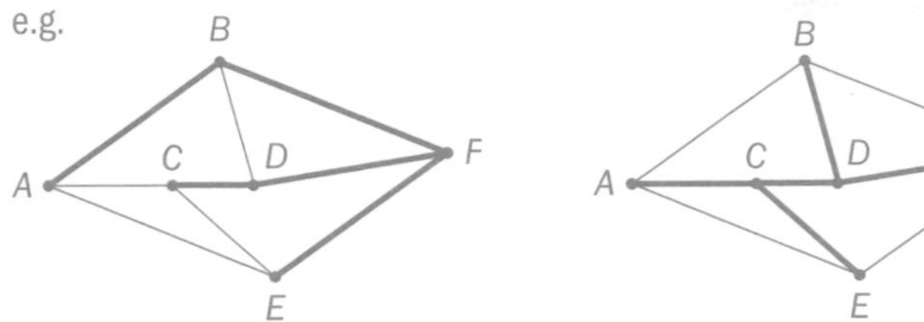

These are two possible spanning trees for the graph shown.

You can see from this example that the spanning tree for a graph with six vertices has five edges. In general, the number of edges is always one less than the number of vertices.

A spanning tree for a graph with n vertices requires $(n - 1)$ edges.

For a network (weighted graph) you can find the spanning tree with the lowest total weight.

For a network, the spanning tree with the lowest total weight is the minimum spanning tree or minimum connector.

e.g. This is the minimum spanning tree for this network.

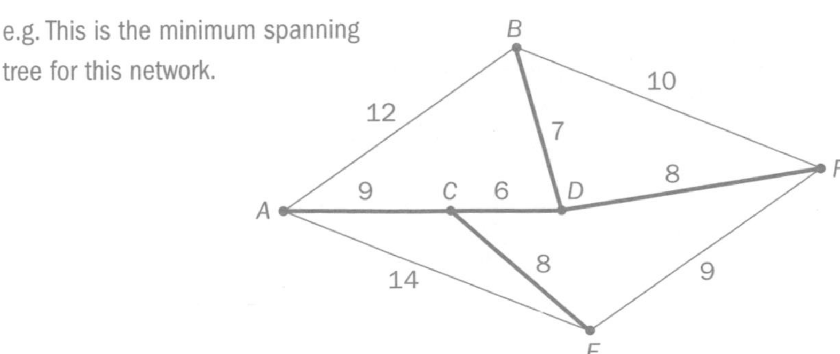

Investigate other possible spanning trees to convince yourself of this.

For a small network like this you can easily find the minimum spanning tree by inspection. However, the number of possible spanning trees increases rapidly as the size of the network increases, so in general you need an algorithm to find the minimum connector. There are two algorithms in common use.

For the complete graph K_n there are n^{n-2} possible spanning trees (this is **Cayley's theorem**). Try drawing the 16 possible spanning trees for K_4.

Kruskal's algorithm

The first method was developed by Martin Kruskal.

Kruskal's algorithm

Step 1 Choose the edge with the minimum weight

Step 2 Choose from the remaining edges the one with the minimum weight, provided that it does not form a cycle with those edges already chosen

Step 3 If any vertices remain unconnected, go to Step 2

If at any stage there is a choice of edges, choose at random.

This is an example of a **greedy algorithm**. At each stage you make the most obviously advantageous choice without thinking ahead. As it turns out, this greedy approach always leads to the minimum connector.

At Step 3, rather than checking for unconnected vertices, you could check whether $(n - 1)$ edges have been chosen.

EXAMPLE 1

Use Kruskal's algorithm to find a minimum spanning tree for the network shown. List the order in which edges are chosen.

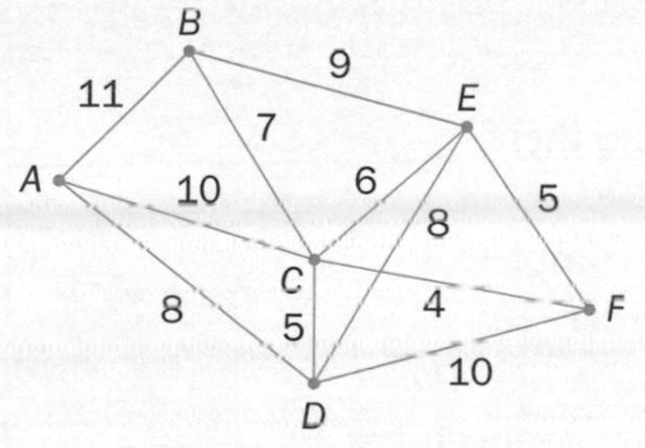

There are 6 vertices, so the connector will have 5 edges.

Choose *CF* first, because this has the least weight.

There is a 'tie' between *CD* and *EF*.
If you choose *CD*, then *EF* will be chosen next.
If you choose *EF*, then *CD* will be chosen next.

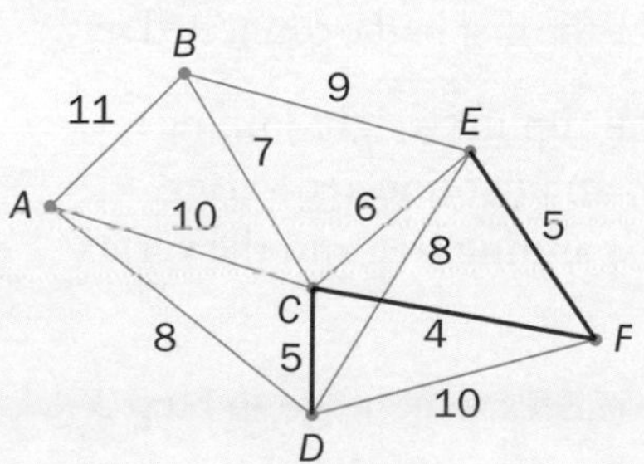

Example 1 is continued on the next page.

D1

EXAMPLE 1 (CONT.)

CE has the next lowest weight but it forms a cycle *CEFC*, so choose *BC* next.

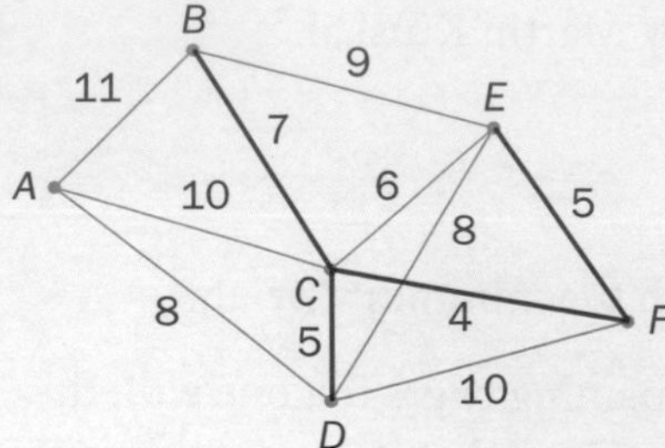

When showing your working in a question you should state when and for what reason you reject a particular edge.

There now seems to be a tie between *AD* and *DE*. But *DE* forms a cycle *DEFCD*, so choose *AD*.

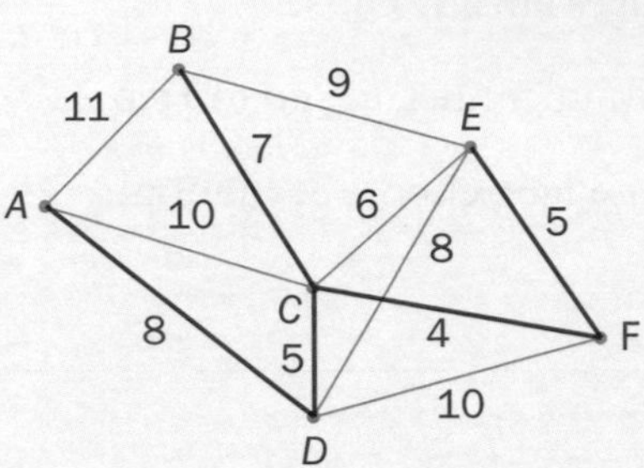

You have now chosen 5 edges and there are no unconnected vertices.
The minimum spanning tree is complete.
The edges were chosen in the order
CF, *CD*, *EF*, *BC*, *AD* or *CF*, *EF*, *CD*, *BC*, *AD*.
The total weight of the minimum connector is 29.

D1

Prim's algorithm

At each stage in Prim's algorithm the vertices are in two sets – the connected and the unconnected vertices.

Prim's algorithm

Step 1 Choose any vertex to be the first in the connected set

Step 2 Choose the edge of minimum weight joining a connected vertex to an unconnected vertex. Add this edge to the spanning tree and the vertex to the connected set

Step 3 If any unconnected vertices remain, go to Step 2

If at any stage there is a choice of edges, choose at random.

This is also a **greedy algorithm**.

Prim's algorithm avoids the problem of forming cycles because it only considers edges to unconnected vertices, which cannot possibly form a cycle.

As before, at Step 3 you could check whether $(n - 1)$ edges have been chosen.

EXAMPLE 2

Use Prim's algorithm to find a minimum spanning tree for the network in Example 1. List the order in which edges are chosen.

Take *A* as the first vertex in the connected set.
The edges to unconnected vertices are *AB*, *AC* and *AD*.
The least weight is *AD*, so this edge is part of the connector.

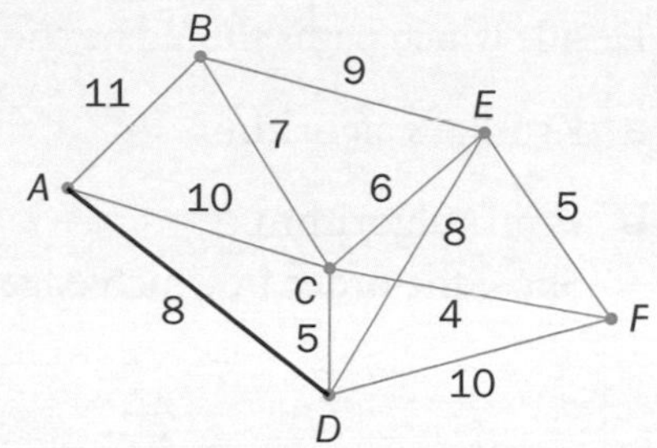

The connected vertices are *A* and *D*.
The edges to unconnected vertices are *AB*, *AC*, *DC*, *DE* and *DF*.
The least weight is *DC*, so this edge is part of the connector.

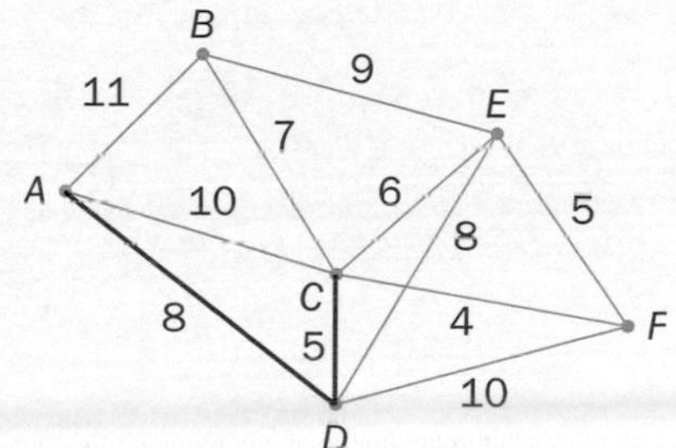

The connected vertices are *A*, *C* and *D*.
The edges to unconnected vertices are *AB*, *CB*, *CE*, *CF*, *DE* and *DF*.
The least weight is *CF*, so this edge is part of the connector.

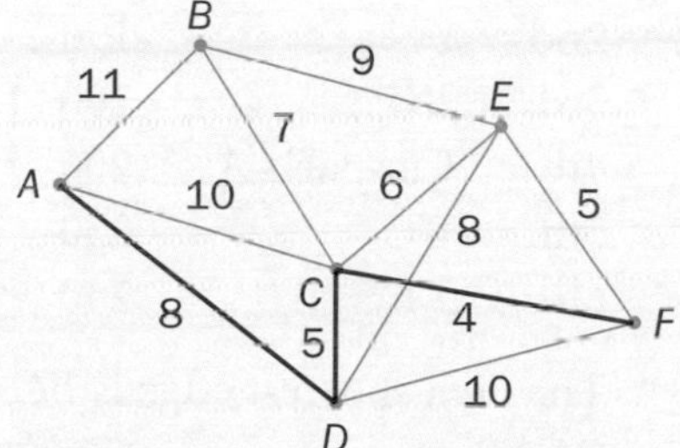

The connected vertices are *A*, *C*, *D* and *F*.
The edges to unconnected vertices are *AB*, *CB*, *CE*, *DE* and *FE*.
The least weight is *FE*, so this edge is part of the connector.

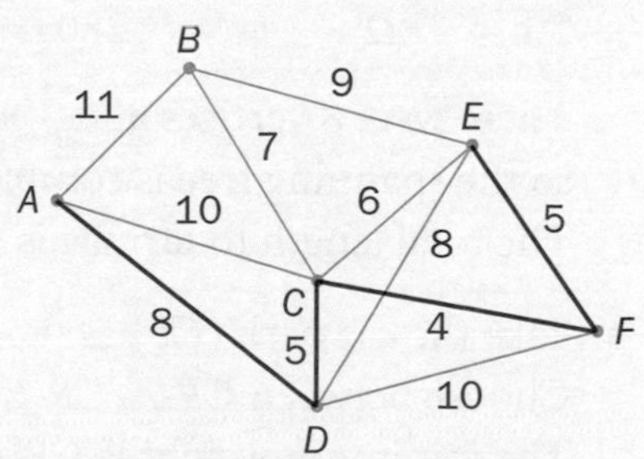

The connected vertices are *A*, *C*, *D*, *E* and *F*.
The edges to unconnected vertices are *AB*, *CB* and *EB*.
The least weight is *CB*, so this edge is part of the connector.

There are no more unconnected vertices, so the minimum spanning tree is complete.

B 11 9 E 7 A 10 6 5 8 C 8 4 F 5 10 D

D1

EXAMPLE 3

The network diagram shows a country park with seven picnic sites and a car park joined by rough paths. Distances are in metres. It is planned to tarmac some of the paths, so that wheelchair users can reach all the picnic sites. Decide which paths should be chosen to minimise the cost using

a Kruskal's algorithm

b Prim's algorithm.
Show the order in which edges are chosen.

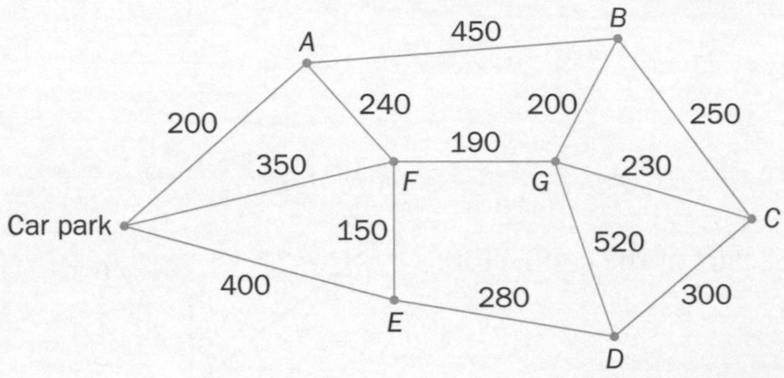

---

a

1st	*EF*	150	(*EF* has the lowest weight)
2nd	*FG*	190	
3rd	*BG*	200	} could be in reverse order
4th	Car park-*A*	200	
5th	*GC*	230	
6th	*AF*	240	

The next shortest edge is *BC*, but this forms a cycle *BGCB*, so

7th	*ED*	280

There were 8 vertices and 7 edges have been chosen, so the spanning tree is complete.
The total length to tarmac is 1490 m.

b Take the car park (*CP*) as the first connected vertex.
Choose between *CP-A*, *CP-E* and *CP-F*.
The minimum weight is *CP-A* = 200.
Add this edge to the spanning tree.

Connected vertices are now *CP* and *A*.
Choose between *CP-E*, *CP-F*, *AF* and *AB*.
The minimum weight is *AF* = 240.
Add this edge to the spanning tree.

Connected vertices are now *CP*, *A* and *F*.
Choose between *CP-E*, *AB*, *FE* and *FG*.
The minimum weight is *FE* = 150.
Add this edge to the spanning tree.

Example 3 is continued on the next page.

EXAMPLE 3 (CONT.)

Continuing in this way, choose $FG = 190$, $BG = 200$, $GC = 230$ and $ED = 280$.

You have chosen 7 edges, so the spanning tree is complete.

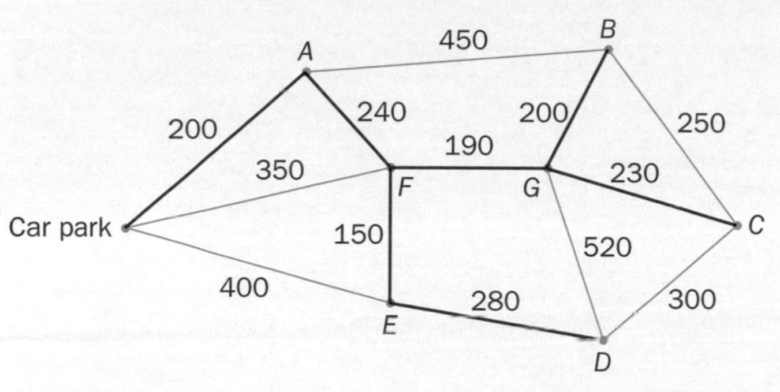

Exercise 3.1

1 Use Kruskal's algorithm to find the minimum connector for the networks shown. List the order in which you choose the edges, and find the total weight of the connector.

a

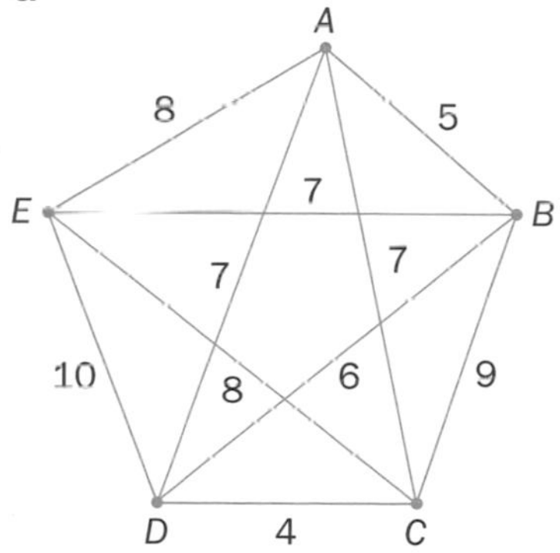

b

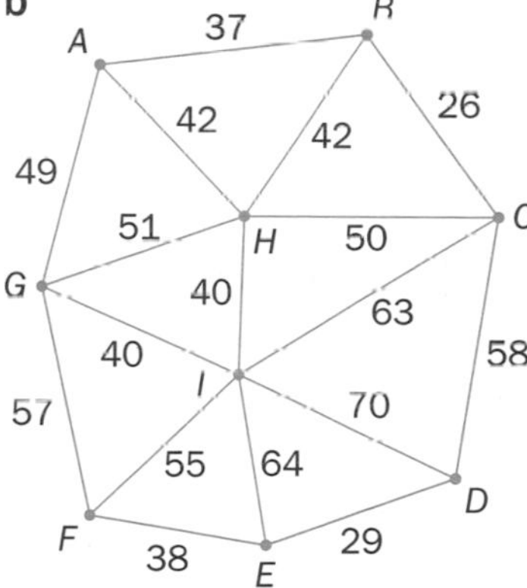

2 Use Prim's algorithm to find the minimum connector for the networks shown in question 1. In each case take A as the starting vertex, and list the order in which you choose the edges.

3 Use Kruskal's algorithm to find the two possible minimum spanning trees for the network shown. List the order in which you choose the edges, and find the total weight of the connector.

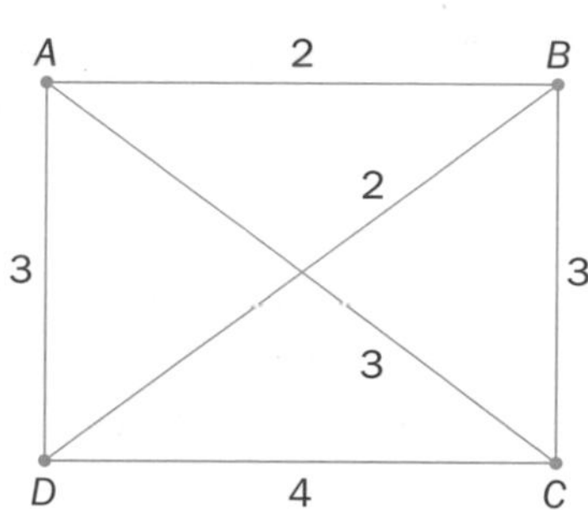

D1

4 Use Prim's algorithm to find the minimum spanning tree for the networks shown. In each case, use A as the starting vertex and list the order in which the vertices are connected. Find the total weight of the spanning tree.

a

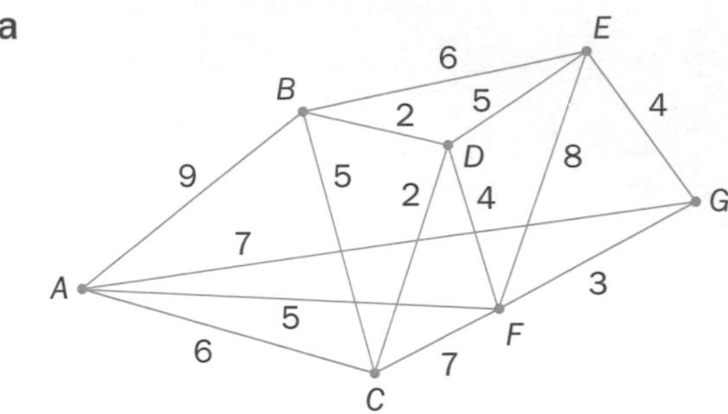

b

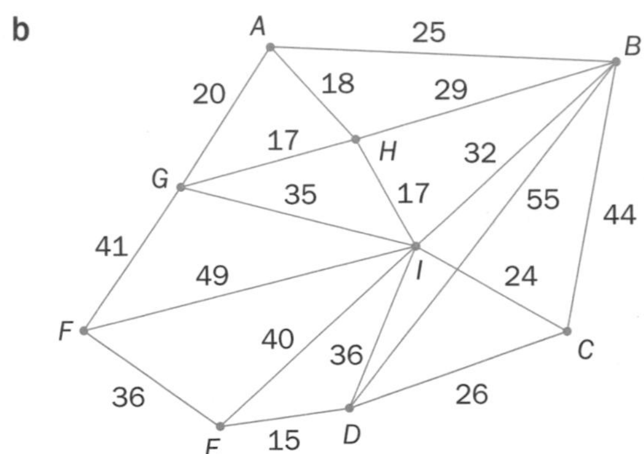

5 A farmer has five animal shelters on his land and wishes to connect them all to the water supply at the farmhouse. The table shows the distances (in metres) between the farmhouse, F, and the shelters A, B, C, D and E (some direct connections are not possible).

	A	B	C	D	E	F
A	—	70	100	120	80	50
B	70	—	—	—	70	—
C	100	—	—	60	—	—
D	120	—	60	—	—	80
E	80	70	—	—	—	—
F	50	—	—	80	—	—

a Draw a network diagram to match the table.

b Use Kruskal's algorithm to find which connections the farmer should make to achieve the water supply as efficiently as possible. List the order in which you choose the connections and find the total length of water piping he would need.

D1

6 The diagram shows the weight limit (in tonnes) for lorry traffic on roads between nine towns. The council plans to ban lorries on as many roads as possible without stopping lorries from reaching all the towns.

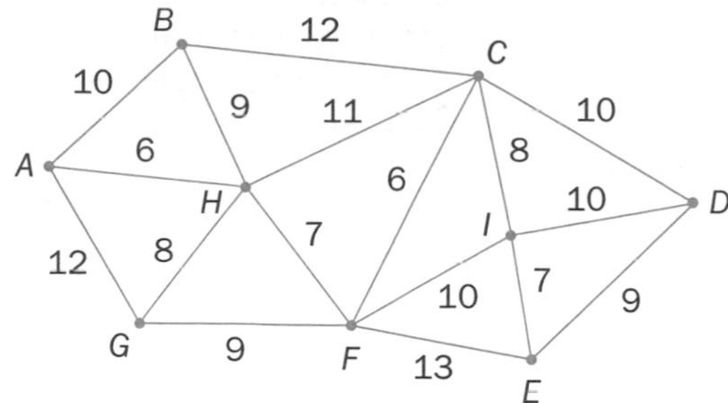

By a suitable modification to Prim's algorithm, choose the roads which should remain open to lorries, and state the heaviest lorry which would then have access to all the towns.

7 The diagram shows the roads connecting five villages, with distances in kilometres.

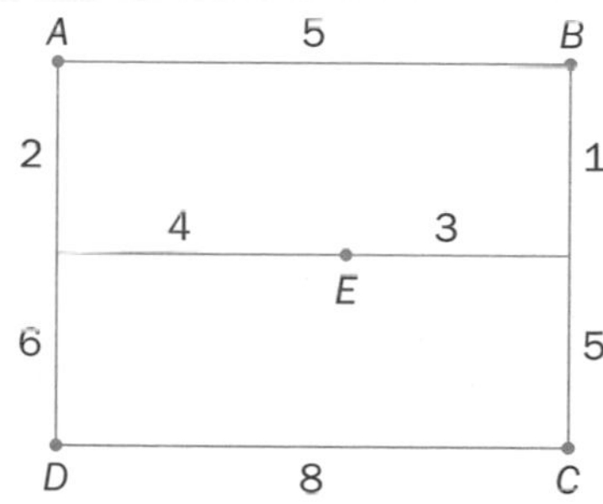

a Construct a table showing the distances by road between the five villages (e.g. the distance from A to E is 6 km).

b Draw a network using the distance matrix from part **a**.

c Use Prim's algorithm, starting from vertex A, to find the minimum connector for the network in part **b**.

As an environmental measure the local council plans to make some stretches of road 'pedestrians, horses and cycles only', leaving just the minimum unrestricted road so that cars can travel between the five villages.

d Explain why the result you obtained in part **c** does not give the number of kilometres of road which must stay open to cars.

e Using the original diagram find, by inspection, the roads which should be kept open to cars.

D1

3.2 Working with a distance matrix

Kruskal's algorithm cannot easily be adapted to a network stored as a matrix.
This means it is not well suited to computerisation.

The problem is that it is difficult to check at each stage whether the edge you are about to choose forms a cycle with those already chosen.

You can however apply Prim's algorithm to a distance matrix.
At each stage you transfer a vertex from the unconnected set to the connected set. On the matrix you

- remove the vertex from the unconnected set by crossing out the row for that vertex
- add it to the connected set by circling the heading of the column for that vertex.

You label the columns 1, 2, 3 etc to show the order in which you chose the vertices.

Prim's algorithm

Step 1 Select the first vertex
Step 2 Cross out the row for the chosen vertex
Step 3 Circle and number the column heading for the chosen vertex
Step 4 Find the minimum undeleted weight in the columns beneath circled vertices. Circle this value. The vertex for this row is the next chosen vertex.
Step 5 Repeat steps 2, 3 and 4 until all vertices have been chosen.

If there is a choice between two equal weights, choose at random.

Look in all the circled columns, not just the most recently circled one.

EXAMPLE 1

Use Prim's algorithm to find the minimum connector for the network given by the distance matrix corresponding to the network on page 56. Use vertex A as the starting vertex.

Cross through row A.
Circle the top of column A and label it 1.

	1					
	(A)	B	C	D	E	F
~~A~~		~~12~~	~~9~~		~~14~~	
B	12	–	–	7	–	10
C	9	–	–	6	8	–
D	–	7	6	–	–	8
E	14	–	8	–	–	9
F	–	10	–	8	9	–

Deleting row A removes A from the unconnected set.

Circling column A adds A to the connected set.

Label 1 because A is the first connected vertex.

Example 1 is continued on the next page.

EXAMPLE 1 (CONT.)

In column A the least weight is for edge AC.

Circle the '9' to show edge AC has been chosen.

Cross through row C.

Circle the top of column C and label it 2.

	1		2			
	A	B	C	D	E	F
A		12	9		14	
B	12	–	–	7	–	10
C	9	–	–	6	8	–
D	–	7	6	–	–	8
E	14	–	8	–	–	9
F	–	10	–	8	9	–

The possible connections are $AB = 12$, $AC = 9$ and $AE = 14$.

Deleting row C removes C from the unconnected set.

Circling column C adds C to the connected set.

Label 2 because C is the second connected vertex.

In columns A and C the least weight is for edge CD.

Circle the '6' to show edge CD has been chosen.

Cross through row D.

Circle the top of column D and label it 3.

	1		2	3		
	A	B	C	D	E	F
A		12	9		14	
B	12	–	–	7	–	10
C	9	–	–	6	8	–
D	–	7	6	–	–	8
E	14	–	8	–	–	9
F	–	10	–	8	9	–

The possible connections are $AB = 12$, $AE = 14$, $CD = 6$ and $CE = 8$.

In columns A, C and D the least weight is DB.

Circle the '7' to show edge DB has been chosen.

Cross through row B.

Circe the top of column B and label it 4.

	1	4	2	3		
	A	B	C	D	E	F
A		12	9		14	
B	12	–	–	7	–	10
C	9	–	–	6	8	–
D	–	7	6	–	–	8
E	14	–	8	–	–	9
F	–	10	–	8	9	–

The possible connections are $AB = 12$, $AE = 14$, $CE = 8$, $DB = 7$ and $DF = 8$.

In columns A, B, C and D the least weight is either CE or DF.

Circle the '8' to show edge CE has been chosen.

Cross through row E.

Circle the top of column E and label it 5.

	1	4	2	3	5	
	A	B	C	D	E	F
A		12	9		14	
B	12			7		10
C	9	–	–	6	8	–
D	–	7	6	–	–	8
E	14	–	8	–	–	9
F	–	10	–	8	9	–

There is a tie between CE and DF, so you choose at random.

In columns A, B, C, D and E the least weight is DF.

Circle the '8' to show edge DF has been chosen.

Cross through row F.

Circle the top of column F and label it 6.

	1	4	2	3	5	6
	A	B	C	D	E	F
A		12	9		14	
B	12	–	–	7	–	10
C	9	–	–	6	8	–
D	–	7	6	–	–	8
E	14	–	8	–	–	9
F	–	10	–	8	9	–

All the vertices are now in the connected set.

The minimum spanning tree is AC, CD, CE, DB and DF, with total weight 38.

Exercise 3.2

1 Use Prim's algorithm to find the minimum spanning tree for the network corresponding to the distance matrix shown.
Use *A* as the starting vertex, listing the order in which the vertices are chosen, and find the total weight of the spanning tree.

	A	B	C	D	E
A	–	5	9	6	3
B	5	–	11	4	7
C	9	11	–	5	8
D	6	4	5	–	12
E	3	7	8	12	–

2 A farmer has five animal shelters on his land and wishes to connect them all to the water supply available at the farmhouse. The table shows the distances (in metres) between the farmhouse, *F*, and the shelters *A*, *B*, *C*, *D* and *E* (some direct connections are not possible).

Use Prim's algorithm to find which connections the farmer should make to achieve the water supply as efficiently as possible. Use the farmhouse as the starting vertex.

	A	B	C	D	E	F
A	–	70	100	120	80	50
B	70	–	–	–	70	–
C	100	–	–	60	–	–
D	120	–	60	–	–	80
E	80	70	–	–	–	–
F	50	–	–	80	–	–

3 a Show by applying Prim's algorithm to the distance matrix shown that the network has two possible minimum spanning trees, and state the total weight.

b Draw network diagrams to illustrate the two possible trees.

	P	Q	R	S	T	U
P	–	2	5	–	–	9
Q	2	–	1	–	4	7
R	5	1	–	4	–	–
S	–	–	4	–	3	6
T	–	4	–	3	–	4
U	9	7	–	6	4	–

4 The table shows the distance by direct rail link between eight towns. During a period of rationalisation it is decided to close some of the links, leaving just enough connections so that it is possible to travel from any town to any other by rail. Use Prim's algorithm to decide which links must be kept so that the amount of track is a minimum. Use town *A* as the starting vertex and record the order in which you make the links.

	A	B	C	D	E	F	G	H
A	–	56	20	–	–	–	–	70
B	56	–	–	15	65	–	75	88
C	20	–	–	87	95	–	120	30
D	–	15	87	–	60	–	25	112
E	–	65	95	60	–	30	40	70
F	–	–	–	–	30	–	45	–
G	–	75	120	25	40	45	–	115
H	70	88	30	112	70	–	115	–

5 The table shows the cost, in pounds per thousand words, of translating between a number of languages.

	English	French	German	Italian	Portuguese	Spanish
English	–	25	30	27	38	22
French	25	–	35	22	36	28
German	30	35	–	35	40	32
Italian	27	22	35	–	26	20
Portuguese	38	36	40	26	–	23
Spanish	22	28	32	20	23	–

a Use Prim's algorithm to find the minimum spanning tree for this network.

b From your solution to part a, state the cheapest sequence of translations to make a document of 10,000 words, originally in German, available in all the languages, and calculate the cost of the operation.

c Calculate the cost of translating the document directly from German into each of the other languages. Suggest reasons why, despite the extra cost, this might be a preferable course of action.

6 A gardener has a patio area paved with square 50 cm slabs. She intends to install a number of small lights, for which she must have a power supply. She plans to hide the cable in the cracks between the paving slabs. Cable can go from light to light, but there are to be no joins in the cable at other points.

The diagram shows the power source at *O* and the lights at *A*, *B*, *C*, *D*, *E* and *F*.

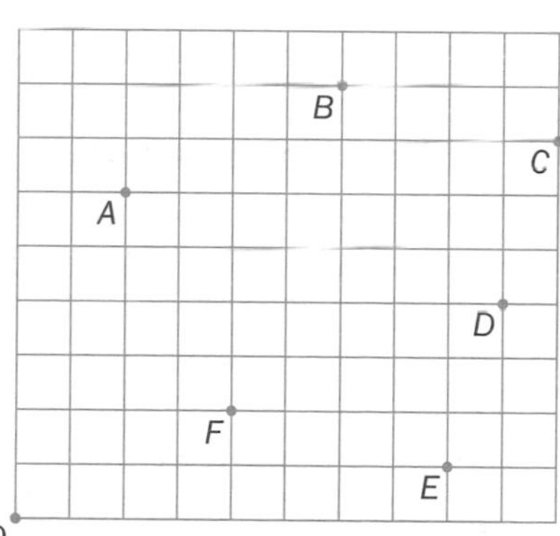

a Ignoring the size of the gaps between the slabs, copy and complete the table for the length of cable run (in metres) between the various lights.

b Use Prim's algorithm, starting from *O*, to find the best layout for the cable. Show all your working and state how much cable will be needed.

c Investigate whether it is possible to improve on this solution if cables may be joined together at other places.

	O	*A*	*B*	*C*	*D*	*E*	*F*
O	–	4	7	8.5	6.5	4.5	3
A	4	–					
B	7		–				
C	8.5			–			
D	6.5				–		
E	4.5					–	
F	3						–

Review 3

1 Use Kruskal's algorithm to find the minimum spanning tree for the graph shown. List the order in which you choose the arcs.

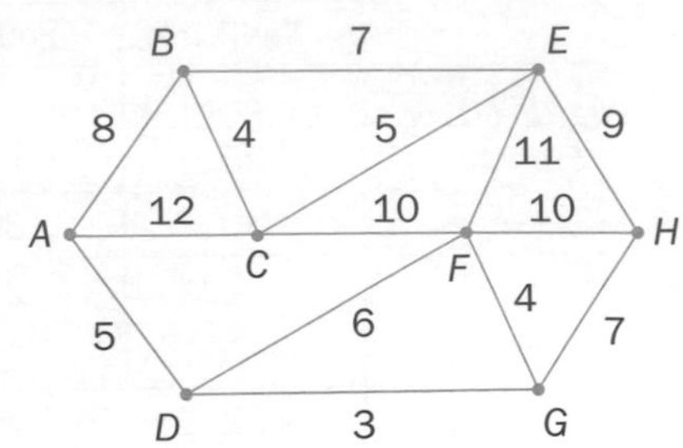

2 Use Prim's algorithm to find the minimum connector for the graph shown. Use *A* as the starting node and list the order in which you choose the arcs.

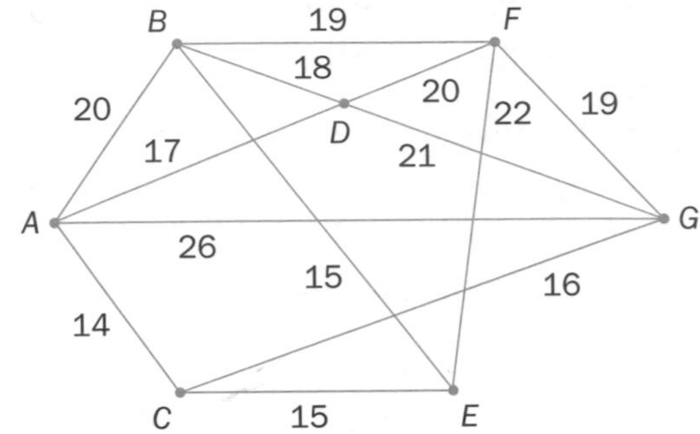

3 Copy the distance matrix shown. Use Prim's algorithm to find the minimum spanning tree for the corresponding network. Use *A* as the starting vertex and indicate at the top of each column the order in which the vertices were chosen. List the order in which the edges were chosen and calculate the total length of the spanning tree.

	A	B	C	D	E	F	G
A	–	9	17	8	22	16	12
B	9	–	6	15	–	20	14
C	17	6	–	–	12	8	19
D	8	15	–	–	10	–	6
E	22	–	12	10	–	11	–
F	16	20	8	–	11	–	13
G	12	14	19	6	–	13	–

4 a Explain why Prim's algorithm is preferred to Kruskal's algorithm when the network is given in the form of a distance matrix.

b The distance matrix shown gives the distances, in km, between six towns. Using Prim's algorithm show that there are two possible minimum spanning trees for the network. In each case draw the arcs which form the spanning tree.

	A	B	C	D	E	F
A	–	4	6	12	15	20
B	4	–	5	8	14	25
C	6	5	–	8	12	24
D	12	8	8	–	19	20
E	15	14	12	19	–	11
F	20	25	24	20	11	–

c A bus company only runs services along the roads in the spanning tree. For each spanning tree

i How far would a person go when travelling by bus from town *A* to town *F*?

ii In which town would you suggest that the company should site the bus station if it wishes to minimise the greatest distance from the station to the towns? Give reasons.

5 The diagram shows the roads connecting seven towns with distances shown in km. During the winter the local council uses snow ploughs to keep enough roads open so that travel is possible between any two towns.

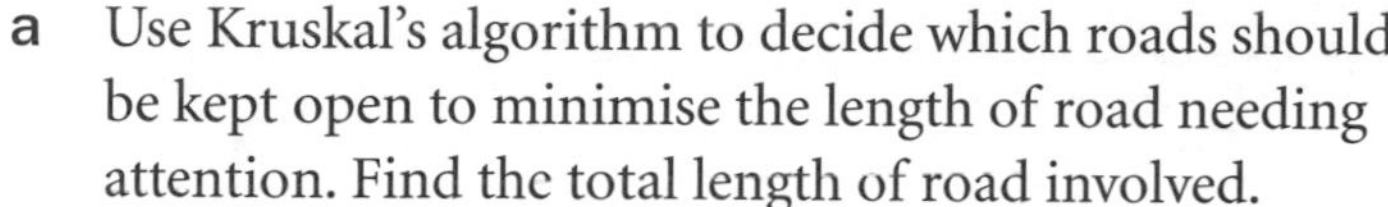

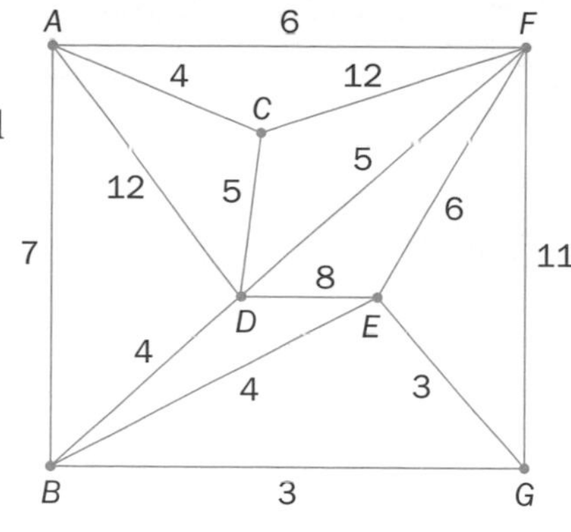

a Use Kruskal's algorithm to decide which roads should be kept open to minimise the length of road needing attention. Find the total length of road involved.

b The ambulance station is situated along the road from A to B, so this must be kept open. Which road would it replace in your answer to part **a**?

6 The table shows the walking times, in minutes, between six tourist attractions in a city. The city information bureau wants to install signposts along a minimum number of routes so that every attraction is accessible and the total walking time is a minimum.

	A	B	C	D	E	F
A	–	4	6	4	3	2
B	4	–	7	6	5	5
C	6	7	–	6	6	5
D	4	6	6	–	3	5
E	3	5	6	3	–	2
F	2	5	5	5	2	–

a Use Prim's algorithm to find which routes it should signpost.

b Draw a tree to represent the minimum connector found in part **a**.

c What is the greatest time a tourist would need to walk from any one attraction to any other?

7 The network shows the distances, in metres, between ten wildlife observation points. The observation points are to be linked by footpaths, to form a network along the arcs indicated, using the least possible total length.

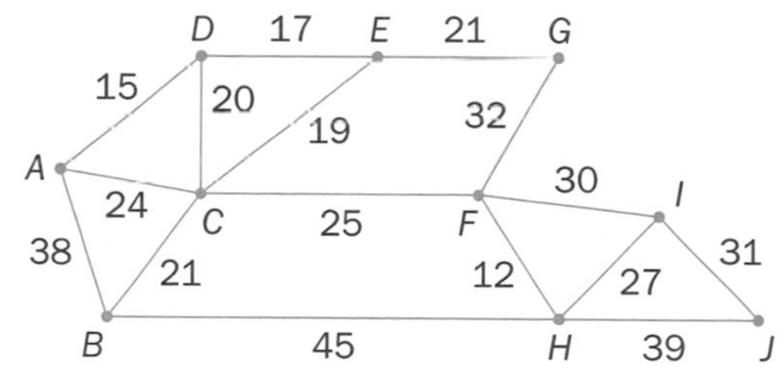

a Find a minimum spanning tree for the network shown, showing clearly the order in which you select the arcs for your tree, using

i Kruskal's algorithm,

ii Prim's algorithm, starting from A.

Given that footpaths are already in place along AB and FI and so should be included in the spanning tree,

b explain which algorithm you would choose to complete the tree, and how it should be adapted. (You do **not** need to find the tree.)

[(c) Edexcel Limited 2005]

3 Exit

Summary

Refer to

- A spanning tree or connector of a graph is a subgraph which is a tree containing every vertex of the graph.
  - A spanning tree for a graph with n vertices requires $(n - 1)$ edges.
  - The minimum spanning tree or minimum connector of a network is the spanning tree with the lowest total weight. 3.1
- Kruskal's algorithm chooses edges in order of their weights, starting with the lowest and avoiding edges which form a cycle with those already chosen. 3.1
- Prim's algorithm chooses a starting vertex and then connects the vertices one at a time. At each stage the next vertex chosen is the unconnected vertex which is 'closest' to a connected vertex. 3.1
- Kruskal's algorithm is difficult to use with a table (distance matrix) and is therefore hard to computerise. Prim's algorithm can be used with a table. 3.2

Links

Finding minimal spanning trees is extremely important for companies that cover a large area. For a national television network using the least amount of cable possible can save them millions of pounds.

4

Shortest path

This chapter will show you how to
- use Dijkstra's algorithm to find the shortest route between two vertices of a network
- understand the limitations on the use of Dijkstra's algorithm.

Introduction

There are many situations in which you need to find the shortest, cheapest or quickest route between two vertices of a network.

e.g. If you enter Southampton and Barnsley into a satellite navigation system, it will calculate the shortest route between these towns. The software makes use of an algorithm to achieve this. One such algorithm is introduced in this chapter.

4.1 Shortest path

In a network, the **shortest path** between two vertices is the one for which the total weight is a minimum.

The weights could, for example, be costs or times rather than distances, so the path might actually be the cheapest or quickest rather than the shortest.

For very small networks you could try out every path, but the number of possible paths increases rapidly as the network size increases, so you need an algorithm which systematically builds the best route.

Dijkstra's algorithm

The usual algorithm is Dijkstra's (*pronounced Dyke-stra's*) algorithm. It is a **labelling algorithm**. You move through the network from the starting vertex, labelling each vertex with a temporary label recording the shortest known distance to that vertex. You update the temporary label if you find a shorter distance, and you make the label permanent when it is clear that you have the shortest possible distance to that vertex.

Each time you pass through the algorithm you give one more vertex a permanent label. The process stops when you give the destination vertex its permanent label.

D1

Dijkstra's algorithm

Step 1 Label the start vertex with permanent label 0

Step 2 If V is the vertex you have just permanently labelled, update those vertices directly connected to V which do not have a permanent label.
For each such vertex X

- Calculate L = label of V + weight of VX
- Give X the temporary label L unless it already has a temporary label less than or equal to L

Step 3 Choose the vertex with the lowest temporary label (choose at random if there is a tie) and permanently label it with this value. If this is the destination vertex then stop. Otherwise go to Step 2

Once you have given the destination vertex its permanent label, you need to trace back to find the route, as follows.

If vertex Y is on the route, you find vertex X such that
(label Y – label X) = weight of XY
Vertex X is the previous vertex on the route

EXAMPLE 1

Find the shortest path from S to T for the network shown.

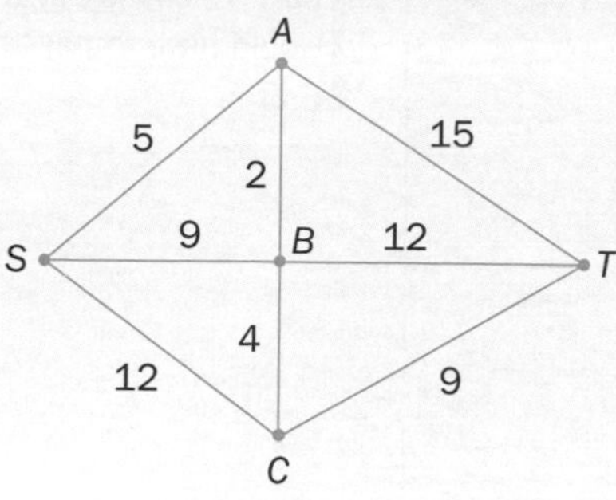

It is common to use S for the start vertex and T for the finish (or terminal) vertex.

Step 1 Give S a permanent label of 0.

Step 2 Give A, B and C temporary labels showing their distance from S.

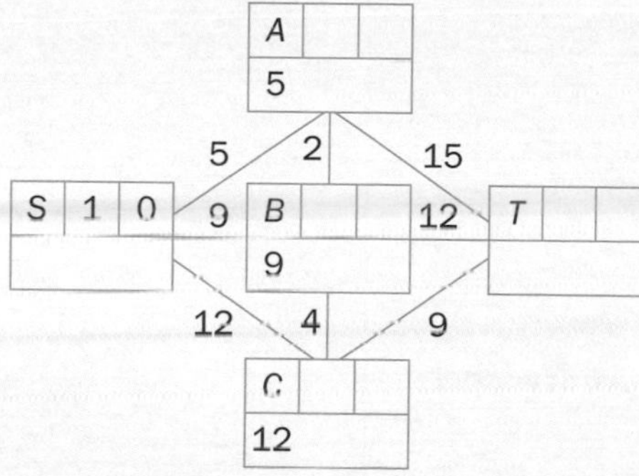

The working is shown in boxes like this:

Vertex	Order of labelling	Final value (permanent label)
Working values (temporary labels)		

This is the layout you should use in the examination. You record the order in which you give the permanent labels so that the examiner can tell that you have correctly followed the algorithm.

Step 3 A has the lowest temporary label. Make this permanent.

The shortest distance to A must be 5, because any other route would have to pass through B or C.

Step 2 The vertices connected to A are T and B.

Calculate for T: $5 + 15 = 20$

T has no label, so give it a temporary label of 20.

for B: $5 + 2 = 7$

This is an improvement, so change the temporary label of B to 7.

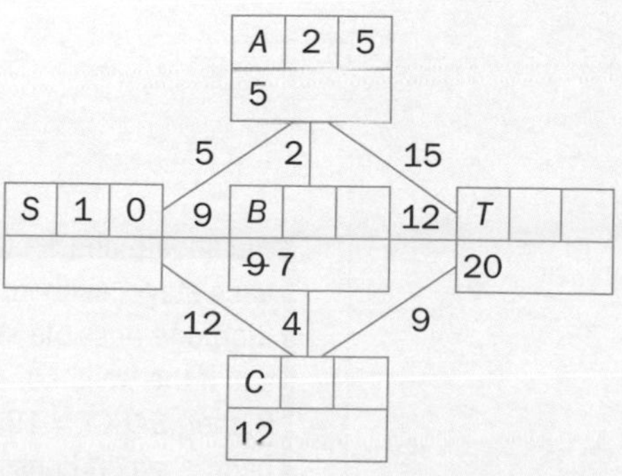

Example 1 is continued on the next page.

D1

EXAMPLE 1 (CONT.)

Step 3 B has the lowest temporary label.
Make this permanent.

Step 2 The vertices connected to B are T and C.
Calculate for T: $7 + 12 = 19$
for C: $7 + 4 = 11$
These are both improvements, so change the temporary labels of T to 19 and of C to 11.

The shortest distance to B must be 7, because any other route would have to pass through C or T.

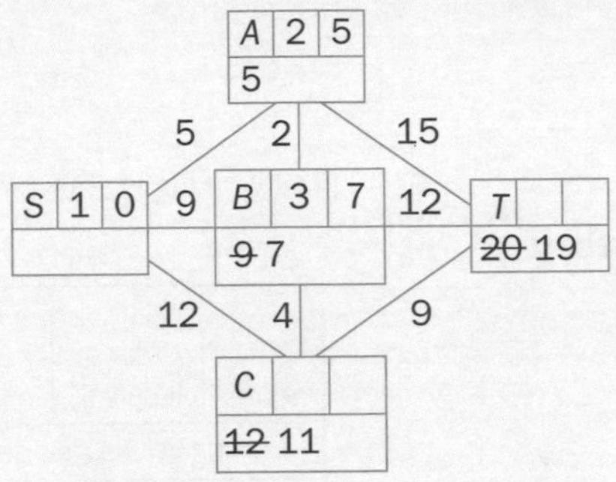

Step 3 C has the lowest temporary label.
Make this permanent.

Step 2 The only vertex connected to C is T.
Calculate for T: $11 + 9 = 20$
This is not an improvement, so the temporary label of T is unchanged.

Step 3 T now has the lowest temporary label.
Make this permanent.
T is the destination vertex, so stop.

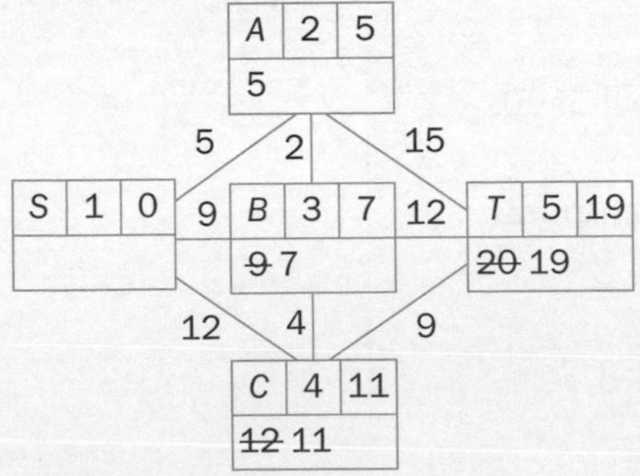

Notice that Dijkstra's algorithm gives the shortest route to **every vertex** which has been permanently labelled.

You now need to trace back and find the route.
Label T – label $B = 19 - 7 = 12$ = weight of BT, so BT is part of the route.
Label B – label $A = 7 - 5 = 2$ = weight of AB, so AB is part of the route.
Label A – label $S = 5 - 0 = 5$ = weight of SA, so SA is part of the route.
The shortest path is therefore $SABT = 19$.

You should check every vertex at each stage, as there may be more than one possible shortest path. e.g. If the weight of CT had been 8, then $SABCT = 19$ would have been a second possible path.

D1

You can use Dijkstra's algorithm with directed networks.

EXAMPLE 2

Find the shortest route between S and T for the directed network shown.

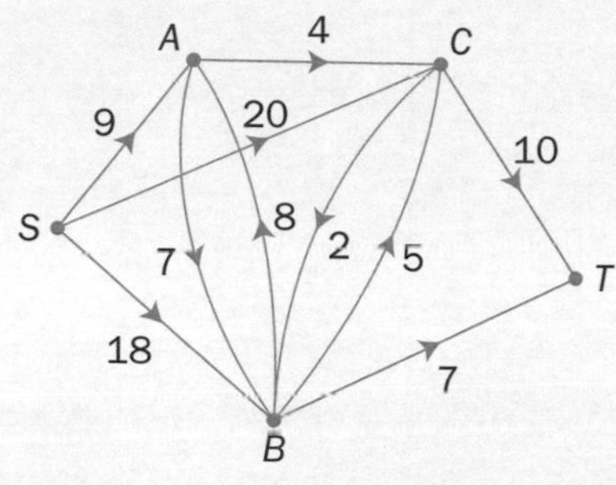

Step 1 Label S with permanent label 0.

Step 2 Give A, B and C temporary labels.

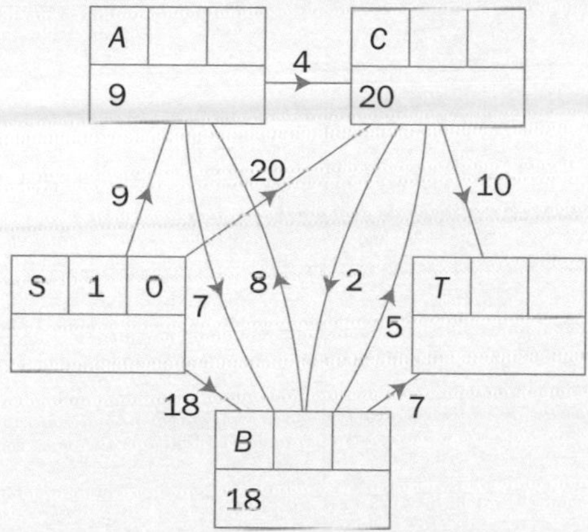

Step 3 A has the lowest temporary label. Make this permanent.

Step 2 Update the temporary labels of B and C:

For B: $9 + 7 = 16$ (improvement)

For C: $9 + 4 = 13$ (improvement)

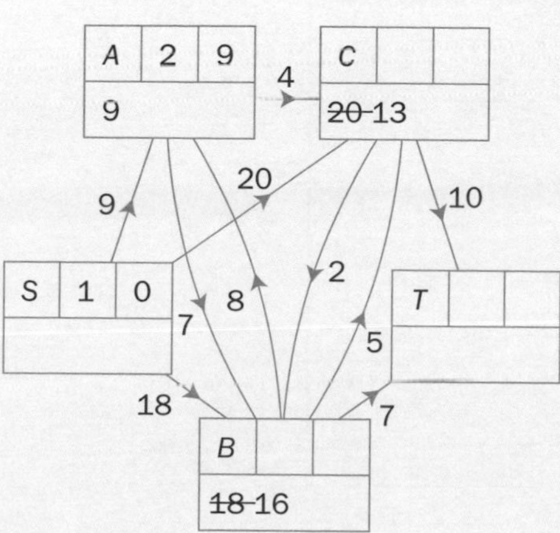

Example 2 is continued on the next page.

D1

EXAMPLE 2 (CONT.)

Step 3 C has the lowest temporary label.
Make this permanent.

Step 2 Update the temporary labels of B and T:

For B: $13 + 2 = 15$ (improvement)

For T: $13 + 10 = 23$ (new label)

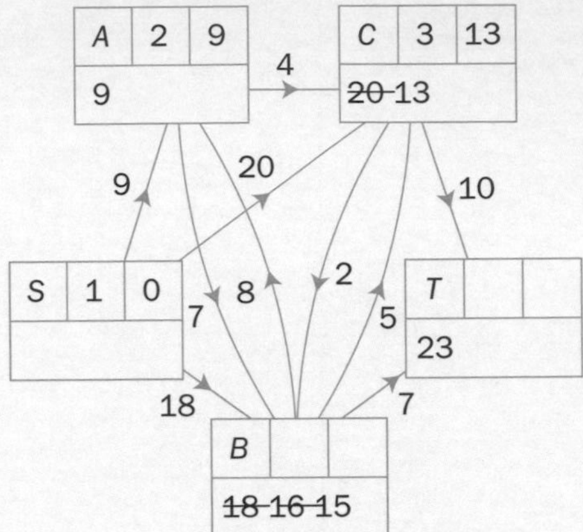

Step 3 B has the lowest temporary label.
Make this permanent.

Step 2 Update the temporary label of T:

For T: $15 + 7 = 22$ (improvement)

Step 3 T has the lowest temporary label.
Make this permanent.
T is the destination vertex. Stop.

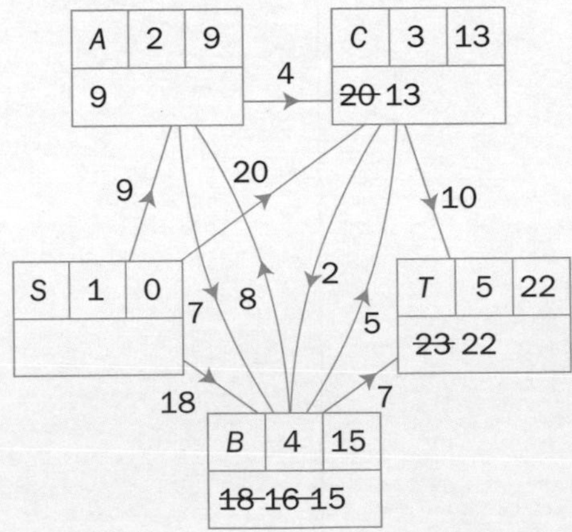

Trace back: Label T – label $B = 7$ = weight of BT
Label B – label $C = 2$ = weight of CB
Label C – label $A = 4$ = weight of AC
Label A – label $S = 9$ = weight of SA

The shortest path is $SACBT = 22$.

D1

Limitation of Dijkstra's algorithm

You cannot use Dijkstra's algorithm if any of the weights are negative.

Negative weights can arise in a number of ways.
e.g. Each weight might represent the cost of travelling along that edge. An edge for which the journey made a profit would appear on that edge as a negative cost.

e.g. Consider this part-network.

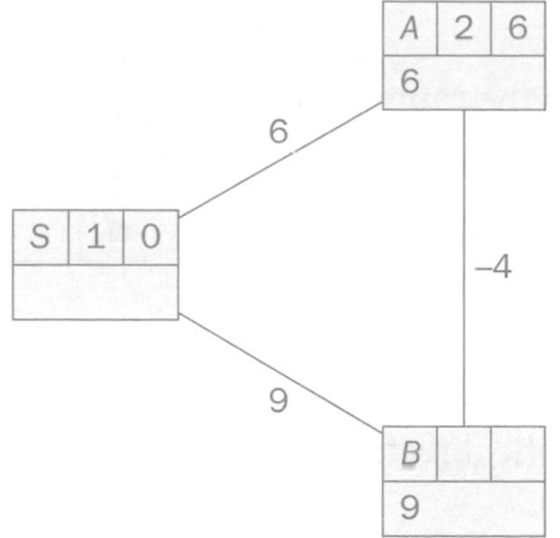

Applying Dijkstra's algorithm, you would give *A* a permanent label of 6, as shown, because at that stage *A* has the lowest temporary label. This permanent label should be the cheapest route from *S* to *A*. However, the route *SBA* would give a total weight of $9 + (-4) = 5$, so Dijkstra's algorithm has failed to find the cheapest route from *S* to *A*.

Exercise 4.1

1 Use Dijkstra's algorithm to find the shortest path from *S* to *T* on this network. Show all your working and state the route and the distance.

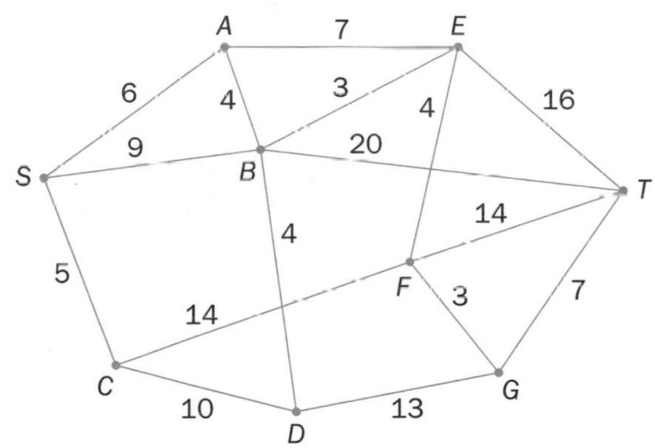

2 Use Dijkstra's algorithm to show that there are two possible shortest routes from *S* to *T* on this network. State the distance and the two possible routes.

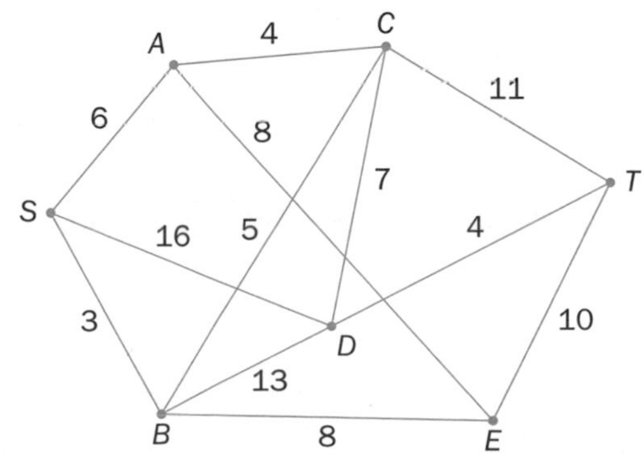

3 The table shows the journey times, in minutes, by bus between seven towns.

   a Draw a network diagram to correspond to this matrix.

   b Use Dijkstra's algorithm to find the route from A to G with the lowest total travelling time.

   c Give reasons why this might not actually be the best route to take.

	A	B	C	D	E	F	G
A	–	30	40	18	–	–	–
B	30	–	8	10	40	–	–
C	40	8	–	25	–	15	–
D	18	10	25	–	45	40	–
E	–	40	–	45	–	10	10
F	–	–	15	40	10	–	35
G	–	–	–	–	10	35	–

4 Use Dijkstra's algorithm to find the shortest route from A to G, and from G to A, for the directed network shown.

5 A distribution company has three warehouses at towns A, B and C. A customer in town L orders an item, which could be supplied from any of the warehouses. The network shows the distances (in kilometres). Use Dijkstra's algorithm to decide which of the three warehouses should supply the item in order to minimise the delivery distance. State the distance and the route taken.

It may appear that you need to apply Dijkstra's algorithm three times to solve this problem. However, if you give some thought to your best strategy, you should be able to answer the question by using the algorithm once only.

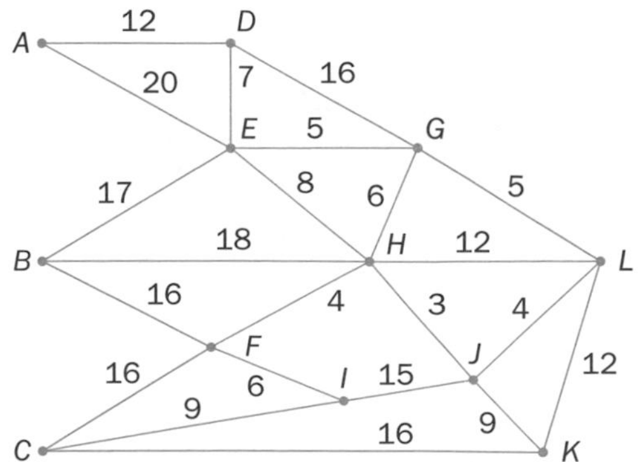

6 The network shows the cost of air travel (in £) between six locations.

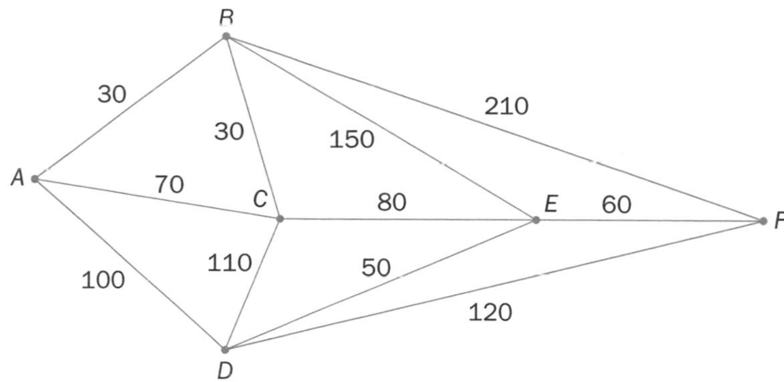

a Use Dijkstra's algorithm to find the cheapest route from A to F.

b A new airport duty is introduced so that every intermediate stop is charged at £20. Apply Dijkstra's algorithm again, adjusting your temporary labels to incorporate this extra cost. What is now the cheapest route?

7 The network shows the weight restrictions (in tonnes) on heavy goods vehicles on a number of roads.

Adapting Dijkstra's algorithm by labelling vertices with the heaviest vehicle which can legally travel to that point, find the heaviest vehicle which can make the journey from A to H, and state the route it should take.

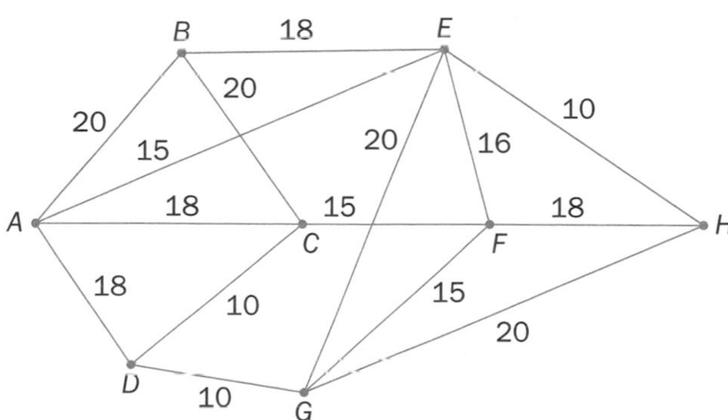

D1

1 The network shown represents the distances in kilometres between eight villages.

Use Dijkstra's algorithm to find the shortest route from S to T. Show all your working.

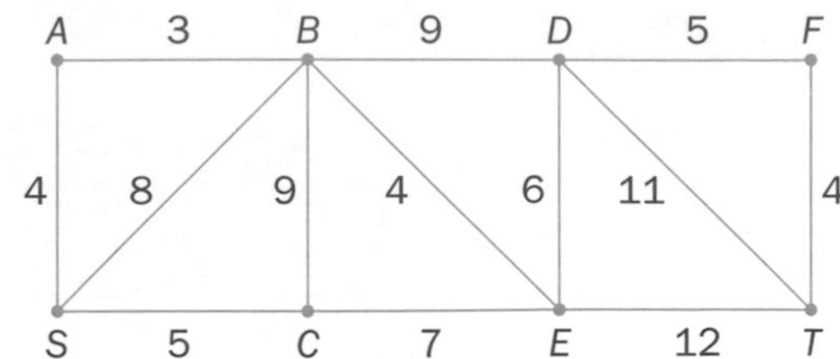

2 The network shows the journey times, in minutes, between eight villages.

a Use Dijkstra's algorithm to find the quickest route from A to D.

b At busy times of day it takes an additional minute to travel through each village en route (except for A and D). Examine whether this alters the quickest route from that found in part **a**.

c Sally travels from A to D at a quiet time of day, but knows that there are roadworks at H which will delay her by 4 minutes. Find her quickest route.

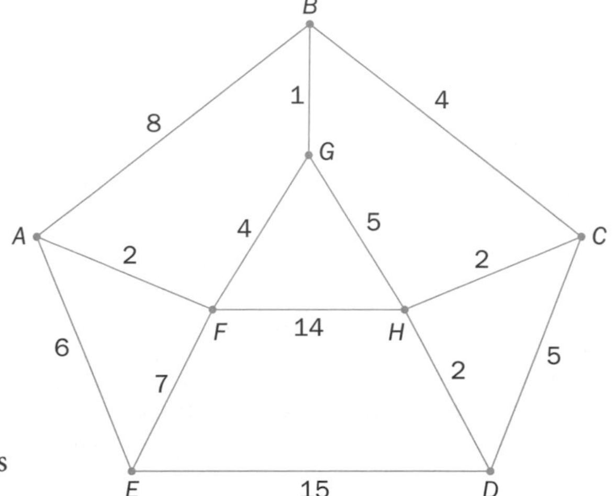

3

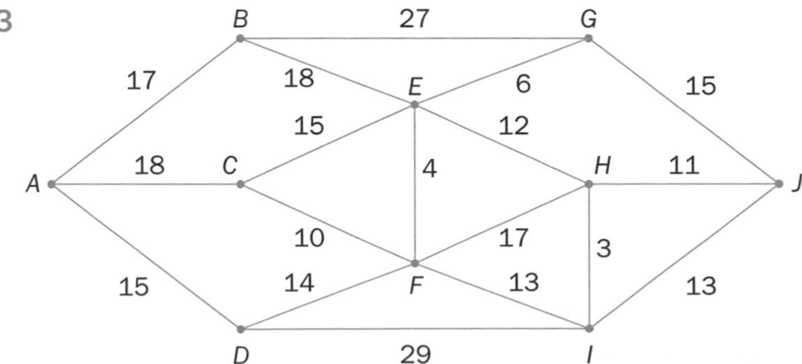

The diagram shows a network of roads. The number on each arc represents the length of that road in km.

a Use Dijkstra's algorithm to find the shortest route from A to J. State your shortest route and its length.

b Explain how you determined the shortest route from your labelled diagram.

The road from C to F will be closed next week for repairs.

c Find the shortest route from A to J that does not include CF and state its length.

4 The diagram shows a network of roads. The weight on each arc represents the cost, in £, of travelling along that road.

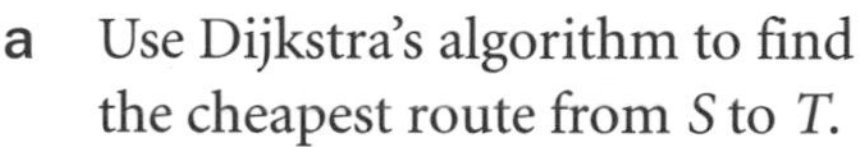

a Use Dijkstra's algorithm to find the cheapest route from S to T.

b On a particular day, Sheila wants to travel from S to T but needs to visit F on the way. By using F as the starting node, use Dijkstra's algorithm to find the cheapest route from S to T via F.

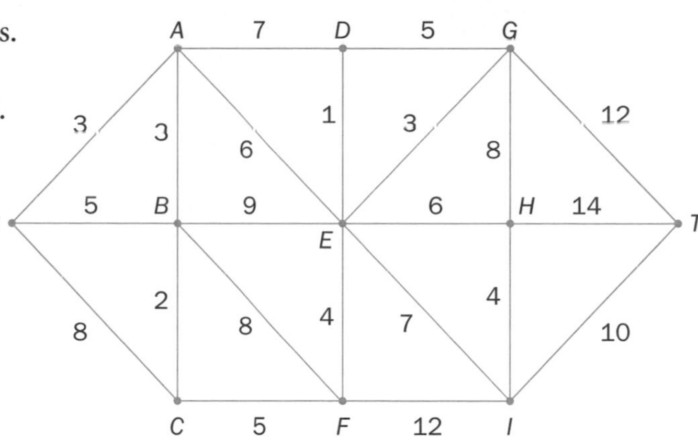

5 The diagram shows a network of roads. The weight on each arc represents the number of traffic lights along that stretch of road.

a Using Dijkstra's algorithm, show that there are two possible routes from P to W with the minimal number of traffic lights.

b On a particular day the road from U to W is closed. Find a route with the minimum number of traffic lights under those circumstances.

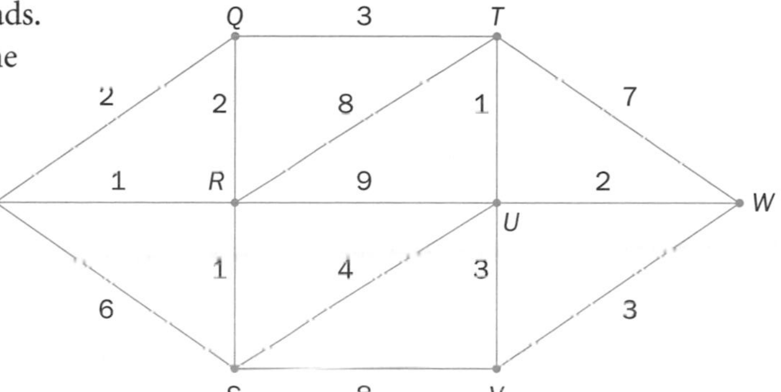

6

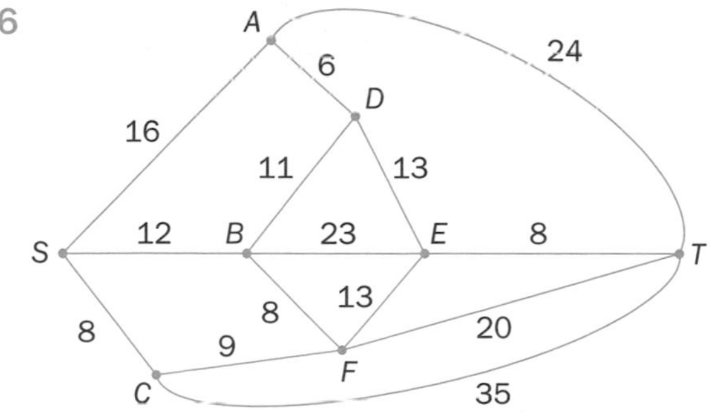

The weighted network shown models the area in which Bill lives. Each vertex represents a town. The edges represent the roads between the towns. The weights are the lengths, in km, of the roads.

a Use Dijkstra's algorithm to find the shortest route from Bill's home at S to T. Show your working on a labelled diagram and explain clearly how you determined the path of least weight from your labelling.

Bill decides that on the way to T he must visit a shop in town E.

b Obtain his shortest route now, giving its length and explaining your method clearly.

[(c) Edexcel Limited 2001]

4 Exit

Summary

Refer to

- The shortest path between two vertices is the one for which the total weight is a minimum. 4.1
- You find the shortest path using Dijkstra's algorithm. This successively labels vertices with their shortest distance from the starting vertex until the destination vertex is labelled. 4.1
- Once the vertices are labelled, you trace back to find the route. An edge XY is on the route if Y is on the route and (label Y – label X) = weight of XY. 4.1
- Dijkstra's algorithm fails if any of the weights are negative. 4.1

Links

As mentioned in this chapter, Dijkstra's algorithm can be used in real life to find the shortest route between two locations. In most cases other factors such as the condition of the landscape also need to be considered. Areas of the world which are prone to landslides are often only accessible by roads. By combining analysis of the land and Dijkstra's algorithm it is possible to develop automatic route planning systems for rugged terrain.

5 Route inspection

This chapter will show you how to

- relate the traversability of a graph (see Section 2.5) to the solution of the route inspection problem for a network
- decide for a given network which edges must be repeated to give the optimum solution to the route inspection problem.

Introduction

In many situations you need to travel at least once along every edge of a network.

e.g. Delivering items to houses along every street in a neighbourhood or inspecting all the cables in a network for faulty insulation.

This chapter looks at the problem of finding the most efficient route in such cases.

The route inspection problem

The route inspection problem can be stated as

For a given network, find the shortest route that travels at least once along every edge and returns to the starting point.

This is sometimes referred to as the **Chinese Postman Problem.** It is the problem, rather than the postman, which is Chinese, having originally been discussed by the Chinese mathematician Mei-ku Kwan.

Ideally you want to travel just once along each edge. You can do this if the network is Eulerian.

See Section 2.5 for more details on Eulerian graphs.

If all vertices in the network are even, the network is traversable. The distance travelled is the sum of the weights.

If there are vertices of odd degree, you will need to repeat some edges to get back to the starting point. You need to choose which edges to repeat to minimise the total distance.

EXAMPLE 1

Find a route passing at least once along each edge of this network, starting and finishing at *A*. Which edge(s) must be repeated in order to achieve this in the minimum distance?

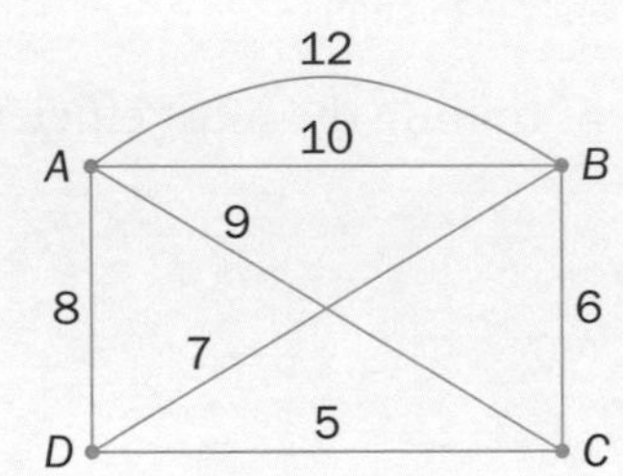

C and *D* are odd vertices, so the network is semi-traversable. If you are to return to your starting point, you need to travel twice between *C* and *D*.
For the minimum total distance, the repeated trip must be by the shortest route, which in this case is along the edge *CD*.

You can modify the diagram to show this.

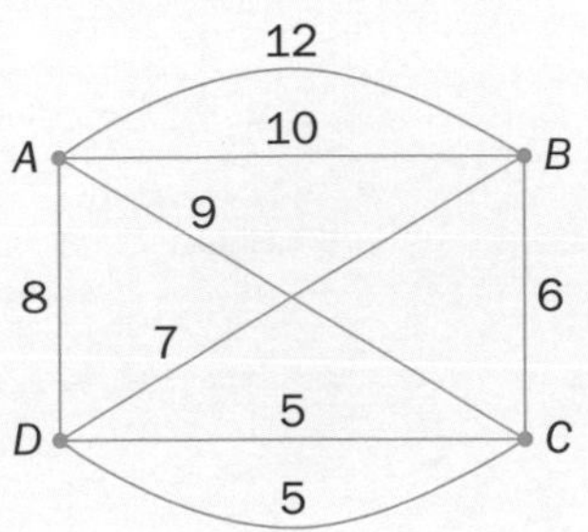

Repeating an edge is equivalent to adding an extra edge to the network diagram.

The modified network is traversable. Any Eulerian trail, such as *ABCDABDCA*, is a possible route.

Total distance = (sum of original weights) + (weight of extra edge)
= 57 + 5 = 62

Example 2 repeats Example 1 but with a different weight on edge *CD*.

EXAMPLE 2

Solve the route inspection problem for the network shown. Find a route starting and finishing at A. Which edge(s) must be repeated in order to achieve this in the minimum distance?

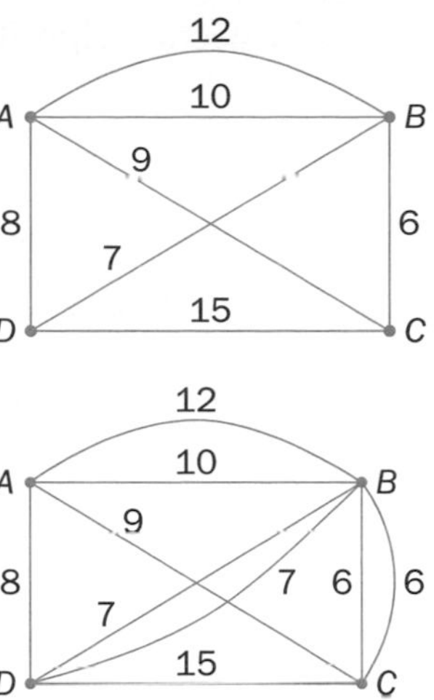

As before, you need to travel twice between C and D.
For the minimum total distance, the repeated trip must be by the shortest route, which in this case is $CBD = 13$.

The modified network is traversable. Any Eulerian trail, such as $ABCDABDBCA$, is a possible route.

Total distance = (sum of original weights) + (weight of extra edges)
$= 67 + 13 = 78$

If there are more than two odd vertices, you can pair them together in various ways. You look for the best way to pair them so that the extra edges have the lowest possible total weight.

Remember there will always be an even number of odd vertices.

EXAMPLE 3

Solve the route inspection problem for this network.

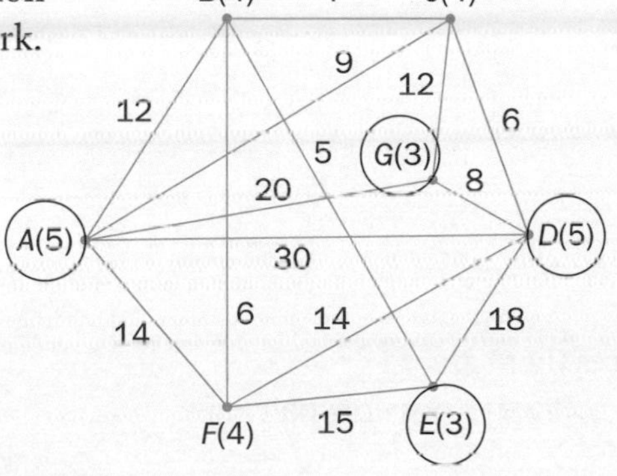

The odd vertices (circled) are A, D, E and G.
You can pair these in three ways:

AD Shortest route $ACD = 15$
EG Shortest route $EBCG = 24$
Total of extra edges = $15 + 24 = 39$

AE Shortest route $ABE = 17$
DG Shortest route $DG = 8$
Total of extra edges = $17 + 8 = 25$

AG Shortest route $AG = 20$
DE Shortest route $DE = 18$
Total of extra edges = $20 + 18 = 38$

The best pairing is AE and DG.
The extra edges are AB, BE and DG.
The total distance is
(sum of weights) + (weight of extra edges) = $176 + 25 = 201$
A possible route, starting from A, is $ABCDEFABEBFDGCAGDA$.

In a complex problem you would use Dijkstra's algorithm to find the shortest routes between the odd vertices, but in small problems it is OK to find the routes by inspection.

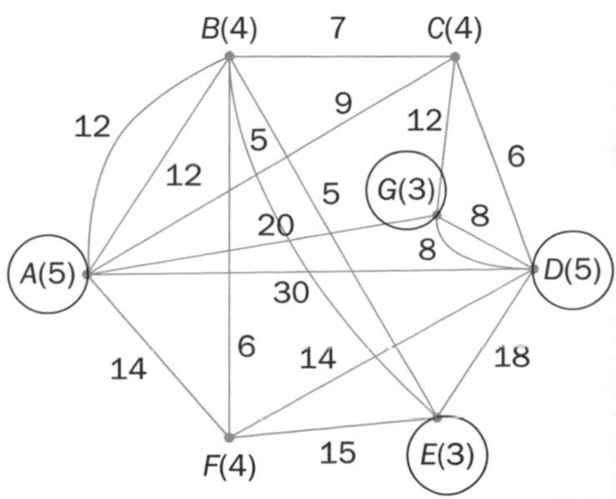

The route inspection algorithm can be stated as

Route inspection algorithm

Step 1 Identify the odd vertices

Step 2 List all possible pairings of the odd vertices

Step 3 For each pairing find the shortest routes between the paired vertices.
Find the sum of the weights involved.

Step 4 Choose the pairing with the smallest sum.
By repeating the shortest route edges (equivalent to adding extra edges to the network), the network becomes traversable.
The total distance = (sum of original weights) + (sum of extra edge weights)

The main difficulty with the route inspection algorithm is that the number of possible pairings increases very rapidly as the number of odd vertices increases.

2 odd vertices – 1 pairing
4 odd vertices – 3 pairings
6 odd vertices – 15 pairings
8 odd vertices – 105 pairings

For a network with n odd vertices there are
$(n-1)\times(n-3)\times(n-5)\times\ldots\times3\times1$ possible pairings.

In the examination you will not be expected to deal with more than four odd vertices.

Example 4 shows how the model used may vary with the practical situation.

EXAMPLE 4

The diagram shows a park, with six gates (one at each vertex) connected by paths of total length 2400 m. A warden patrols the park.

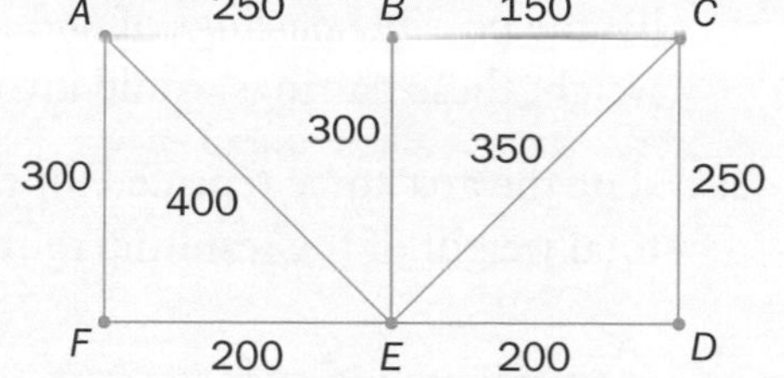

a Find the warden's best route if he enters and leaves the park by gate *A*, and walks along each path at least once.

b Find his best route if he can enter and leave by different gates.

---

a The odd vertices are *A*, *B*, *C* and *E*.
You can pair these in three ways:

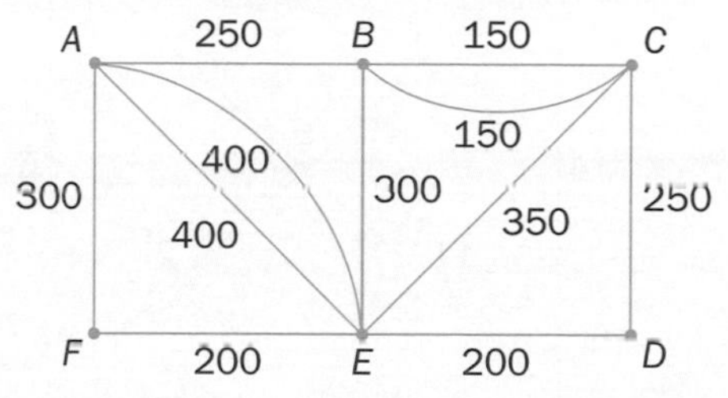

AB/*CE*: total extra weight = 250 + 350 = 600
AC/*BE*: total extra weight = 400 + 300 = 700
AE/*BC*: total extra weight = 400 + 150 = 550

The best route repeats *AE* and *BC*, as shown.
The distance travelled = sum of weights + extra weight
= 2400 + 550 = 2950 m

A possible route is *ABCDEFAECBEA*.

b As the start and end vertices can now be different, you only need the graph to be semi-Eulerian. This means the route must repeat the journey between two of the odd vertices. The shortest extra distance is *BC* = 150, giving the network shown.

The route must enter at *A* and leave at *E* (or vice versa).
The distance = sum of weights + extra weight
= 2400 + 150 = 2550 m

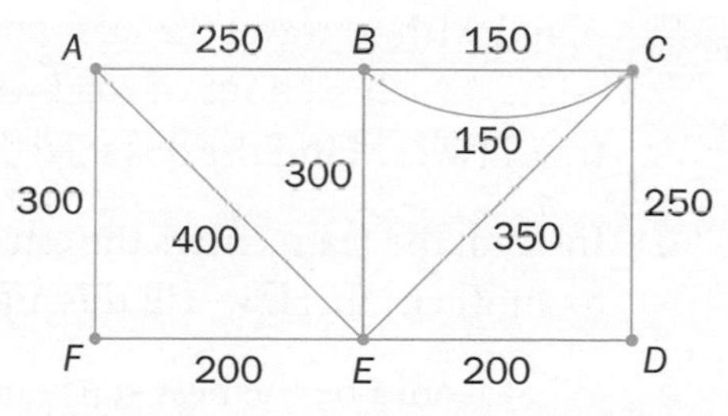

A possible route is *ABCDEFAECBE*.

Exercise 5.1

1 For the network shown

a determine the edges which should be repeated to solve the route inspection problem

b state the length of the route and give a possible route starting from vertex *A*.

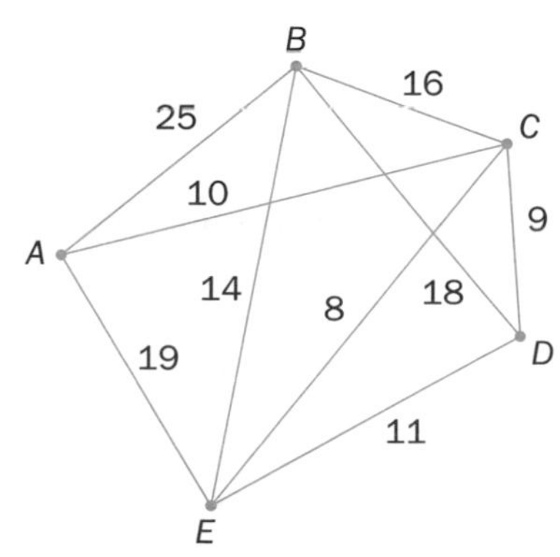

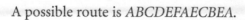

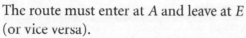

2 Solve the route inspection problem for each of the networks shown. In each case

i list the possible pairings of odd nodes and the total extra weight these pairings would involve

ii state the arcs to be repeated for the best solution and the total weight of the resulting route.

a

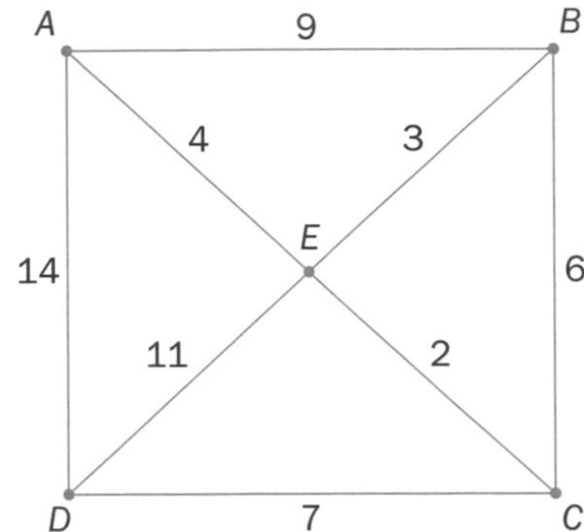

b

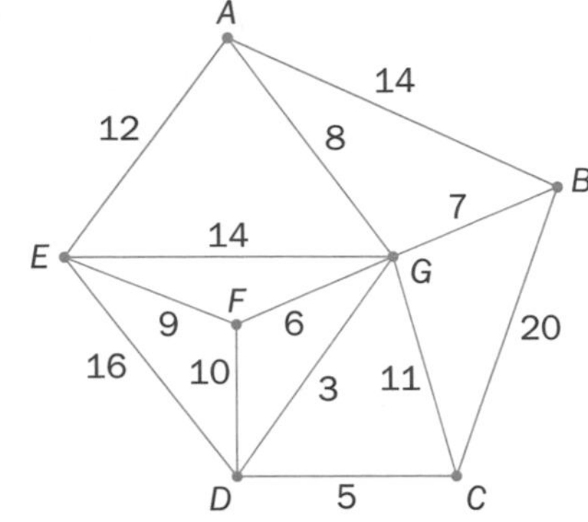

3 The diagram shows the map of the roads on an estate, with distances in metres. All junctions are right angles and all roads are straight apart from the crescent. A team of workers is to paint a white line along the centre of each road.

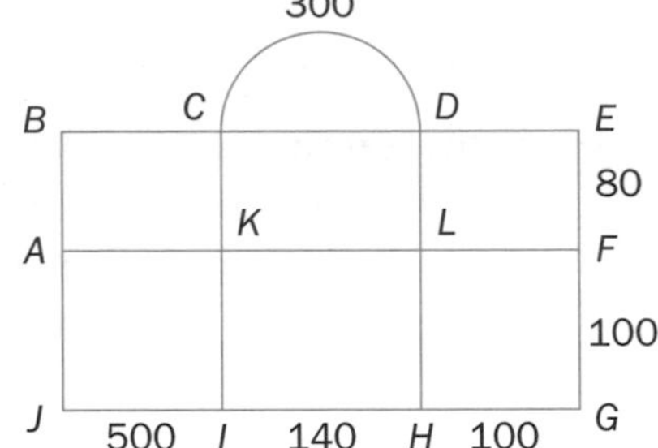

a Assuming the team enter and leave the estate at A and wish to travel the least possible distance on the estate, which sections of road will they will need to travel twice? Find the distance they will travel.

b Instead, the team enters the estate at *H* and leaves at *I* to go on to another job. How will this affect your results from part **a**?

c What would be the best start and end points for the team if they wish to travel the least possible distance? How far would they go in this case?

4 The table shows the direct road links between six towns, with distances in kilometres. Abigail is to do a sponsored cycle ride, travelling at least once along each of these roads and starting and finishing at her home in town *A*.

a Draw the network diagram corresponding to this table.

b Investigate which roads Abigail should repeat and how long her journey will be.

	A	*B*	*C*	*D*	*E*	*F*
A	–	15	–	8	20	–
B	15	–	10	–	6	5
C	–	10	–	9	–	14
D	8	–	9	–	9	4
E	20	6	–	9	–	–
F	–	5	14	4	–	–

5 The diagram shows the layout of paths in a small pedestrian shopping precinct. Distances are in metres.

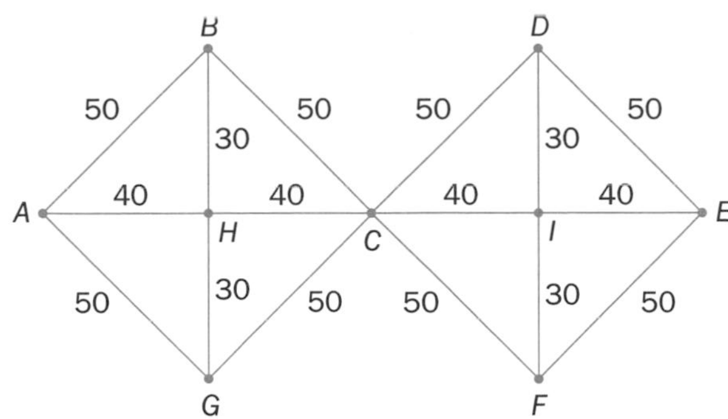

It is necessary to drive a cleaning cart around the precinct, starting and ending at *A*.

Although there are six odd vertices in this problem, there is sufficient symmetry to make the task reasonable.

a Find the sections which should be travelled twice to minimise the distance travelled.

b Find the total distance travelled and suggest a possible route.

6 The diagram shows a logo for a sports goods company, displayed outside a shop.
The lengths shown are in cm and the logo is symmetrical, with all angles right angles.

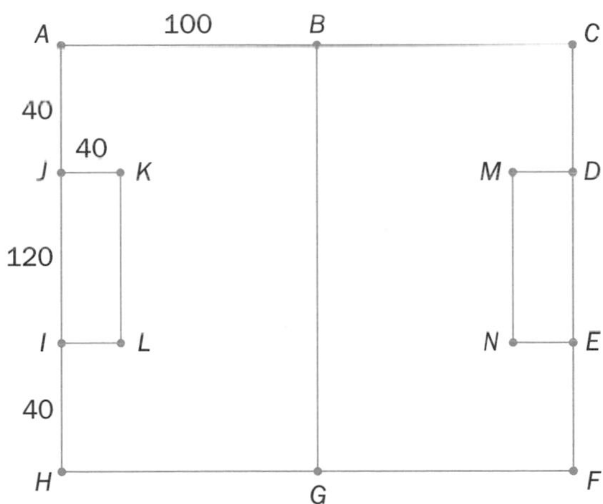

a The shop owner wishes to outline the logo with a cable of flashing lights. She has a single cable which will start and end at *A*. Calculate the minimum length of cable needed, and state which sections of the logo will have a double run of cable.

b If the cable can start and end at different vertices, calculate the minimum length of cable needed. State the start and end points and the section(s) with a double run of cable.

c If she wants to avoid double runs of cable, she will need to use more than one cable. Investigate her best strategy.

1 The diagram shows the roads connecting five towns, labelled with the distances in kilometres. The total of the distances is 86 km.

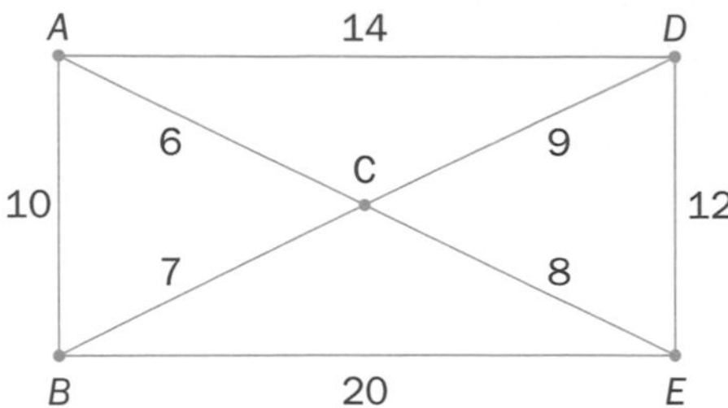

The highway department sends a vehicle from its depot at *A* to inspect all the roads for potholes. The vehicle must return to *A* when the inspection is complete.

a If the vehicle must travel both ways along each road to inspect both carriageways, explain why it can do the job in a minimum distance of 172 km (that is, 2 × 86 km).

b If the vehicle only needs to travel once along a road to adequately inspect it, find the minimum distance it needs to travel, and state which roads must be travelled twice.

2 The diagram shows the paths in a nature reserve, with lengths marked in metres. Barry lost his wedding ring during a visit. Deborah was going there the next day and agreed to look for it.

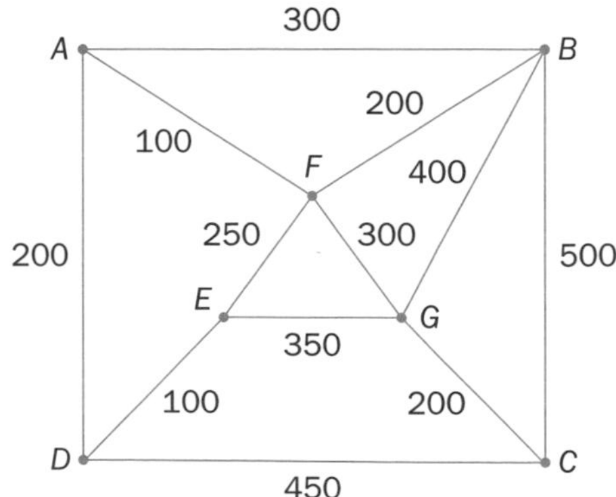

a If Deborah needs to search along every footpath, what is the minimum distance she might need to walk? Which paths will she travel along twice?

b Barry is sure he didn't go along BG. Explain why, nevertheless, Deborah might choose to walk along *BG*.

3 The network represents the paths in a park, with distances in metres. The council buys a robot litter picker, which will be programmed to clean along every path daily.

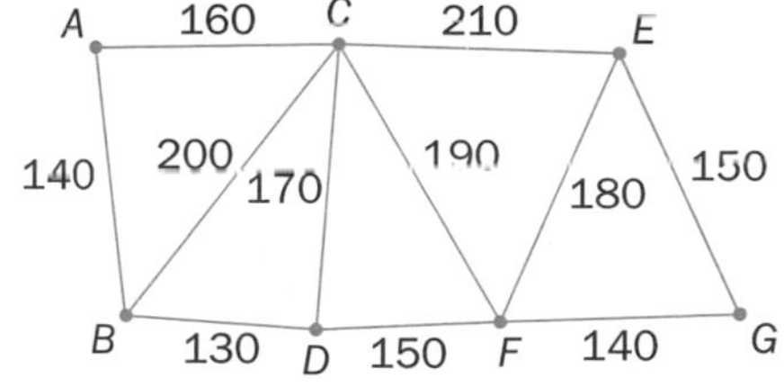

a If the robot has to start and finish each day at the same point

i what is the least distance it will travel during the day?

ii which paths will it travel more than once?

b If the robot can start and finish at different points

i which points should be chosen as the start and finish?

ii how far will the robot travel?

iii which paths will it travel more than once?

4 The diagram models a network of roads which need to be inspected to assess if they need to be resurfaced. The number on each arc represents the length, in km, of that road. Each road must be traversed at least once and the length of the inspection route must be minimised.

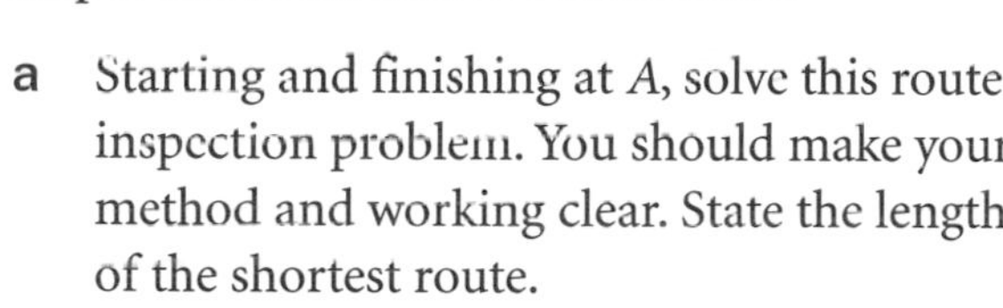

a Starting and finishing at A, solve this route inspection problem. You should make your method and working clear. State the length of the shortest route.
(The weight of the network is 77 km.)

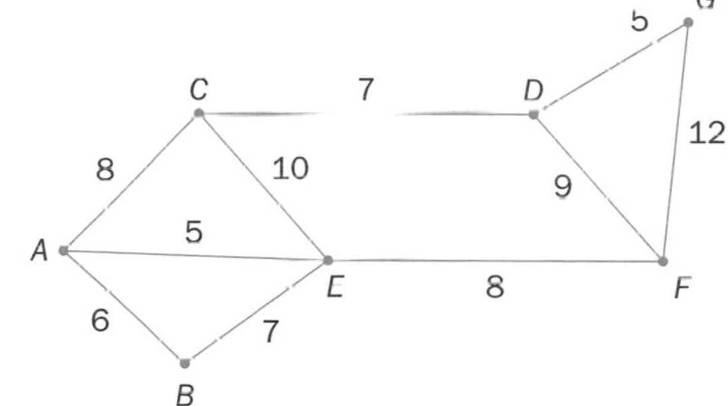

Given that it is now permitted to start and finish the inspection at two distinct vertices,

b state which two vertices you should choose to minimise the length of the route. Give a reason for your answer.

[(c) Edexcel Limited 2005]

5 a Explain why it is impossible to draw a network with exactly three odd vertices.

The route inspection problem is solved for the network shown and the length of the route is found to be 100.

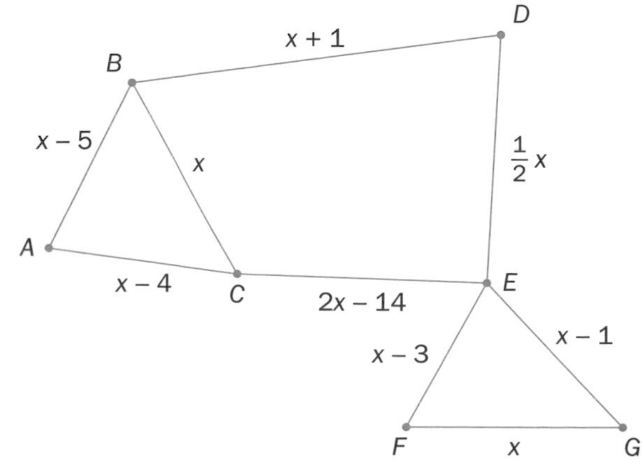

b Determine the value of x, showing your working clearly.

[(c) Edexcel Limited 2003]

5 Exit

Summary

Refer to

- You can solve the route inspection problem (Chinese postman problem) to find a closed route of minimum total weight that passes along every edge of the network at least once.
  - If the network is Eulerian (all vertices are of even degree), any Eulerian trail is a possible route, and the length of the route is the total weight of the network.
  - If the network is non-Eulerian, you find the pairing of odd vertices for which the total of the shortest paths between paired vertices is a minimum.
    By repeating the edges in these shortest paths, the network becomes Eulerian.
    The length of the route = (weight of network) + (weights of repeat edges) 5.1
- In some cases the inspection route can start and end at different vertices. In this case you only need a semi-Eulerian network, with the two odd vertices being the start and end vertices. If there are more than two odd vertices, you need to repeat edges between all but one pair of odd vertices. 5.1

Links

The route inspection problem has many applications in real life. An obvious example is devising routes for rubbish collections.
In addition, the route inspection algorithm is applied in many less obvious situations. It is used when devising test systems for websites. In this case the pages are the vertices of the graph and the links are the edges with the aim being to create a system for testing every link.
Route inspection can even be used in DNA analysis.

1 90, 50, 55, 40, 20, 35, 30, 25, 45

a Use the bubble sort algorithm to sort the list of numbers into descending order showing the rearranged order after each pass.

Jessica wants to record a number of television programmes onto video tapes. Each tape is 2 hours long. The lengths, in minutes, of the programmes she wishes to record are

55, 45, 20, 30, 30, 40, 20, 90, 25, 50, 35 and 35.

b Find the total length of programmes to be recorded and hence determine a lower bound for the number of tapes required.

c Use the first fit decreasing algorithm to fit the programmes onto her 2-hour tapes.

Jessica's friend Amy says she can fit all the programmes onto four tapes.

d Show how this is possible. [(c) Edexcel Limited 2001]

2 45, 56, 37, 79, 46, 18, 90, 81, 51

a Using the quick sort algorithm, perform **one** complete iteration towards sorting these numbers into **ascending** order.

b Using the bubble sort algorithm, perform **one** complete pass towards sorting the **original** list into **descending** order.

Another list of numbers, in ascending order, is

7, 23, 31, 37, 41, 44, 50, 62, 71, 73, 94

c Use the binary search algorithm to locate the number 73 in the list. [(c) Edexcel Limited 2004]

3 a Describe a practical problem that could be modelled using this network and solved using the route inspection algorithm.

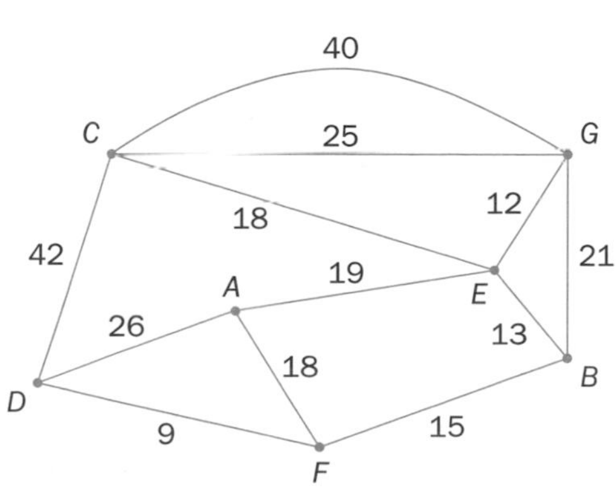

b Use the route inspection algorithm to find which paths, if any, need to be traversed twice.

c State whether your answer to part **b** is unique. Give a reason for your answer.

d Find the length of the shortest inspection route that traverses each arc at least once and starts and finishes at the same vertex.

Given that it is permitted to start and finish the inspection at two distinct vertices,

e find which two vertices should be chosen to minimise the length of the route. Give a reason for your answer. [(c) Edexcel Limited 2004]

4 The diagram describes an algorithm in the form of a flowchart, where a is a positive integer.

List P, which is referred to in the flowchart, comprises the prime numbers 2, 3, 5, 7, 11 , 13, 17, …

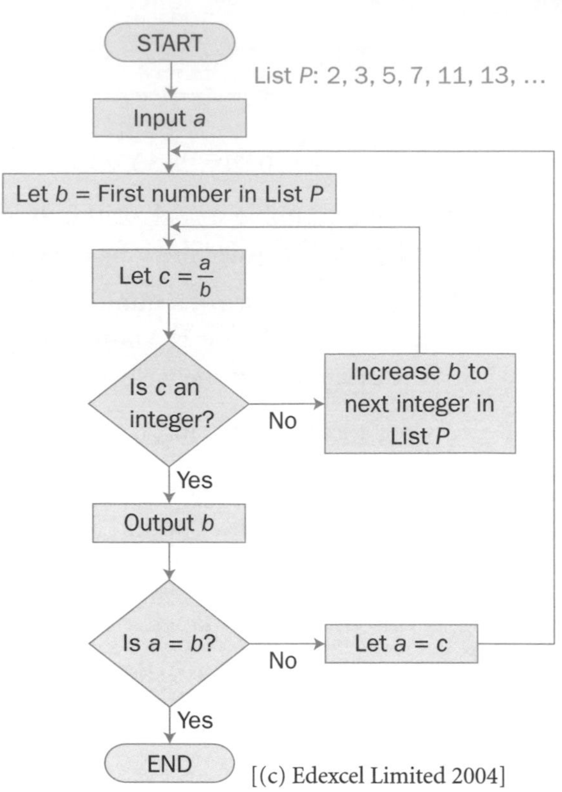

[(c) Edexcel Limited 2004]

a Starting with $a = 90$, implement this algorithm, showing your working in a table like this:

a	b	c	Integers?	Output list	$a = b$?

b Explain the significance of the output list.

c Write down the final value of c for **any** initial value of a.

5 The vertices of this network represent ten towns A–J. The arcs show direct bus routes, and the weights are the journey times in minutes.

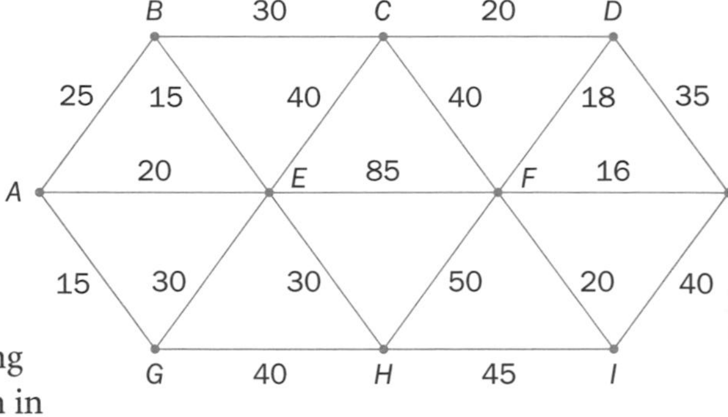

a i Use Dijkstra's algorithm to find the length of the shortest path from A to J. Show your labelling clearly on a copy of the diagram.

ii Hence, assuming no time is lost when changing buses, state the routes which should be chosen in order to minimise the time taken to travel from A to J.

b In fact it is necessary to change buses in each town. Each change takes 10 minutes. Show that the route found in part **a ii** is no longer the quickest way of getting from A to J.

c On Sundays each change takes 20 minutes. Find the quickest way of getting from A to J by bus on a Sunday. Give a reason for your answer.

6 **a** Describe the differences between Prim's algorithm and Kruskal's algorithm for finding a minimum connector of a network.

b Listing the **arcs** in the order that you select them, find the minimum connector for the network in the diagram, using

i Prim's algorithm ii Kruskal's algorithm.

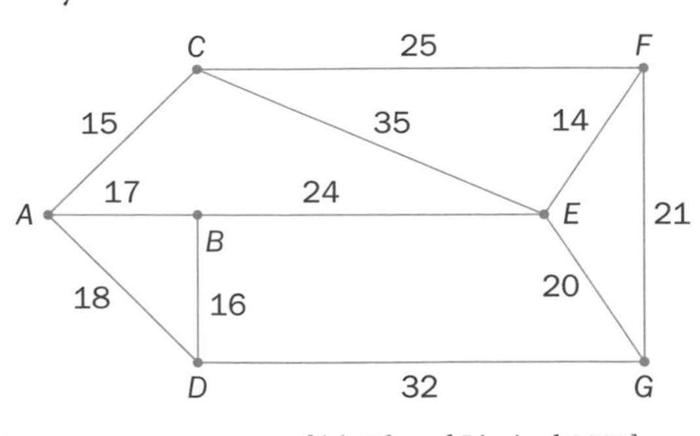

[(c) Edexcel Limited 2003]

6 Matchings

This chapter will show you how to

- model a problem involving matching items from two sets using a bipartite graph
- understand the terms maximal matching and complete matching
- use the maximum matching algorithm to improve on a given matching and to show if a matching is maximal.

Introduction

You saw in Section 2.3 that problems involving pairing up members of two sets, such as allocating workers to tasks for which they are qualified, can be modelled using bipartite graphs.

Chapter 6 examines how you can use a systematic approach to find the best possible allocation in a given situation.

6.1 The matching problem

You can illustrate the possible connections between members of two sets using a bipartite graph. The next stage is to choose which of those connections give the best match between the sets.

EXAMPLE 1

A school runs three language classes, one each of French, German and Spanish. There are four language teachers – Polly Glott, who can speak all three languages, Lynne Gwist, who speaks German and Spanish, Frank O'File, who speaks just French, and Dick Shonary, who speaks French and Spanish. Find an optimum allocation of teachers to classes.

The skills of the four teachers are shown by this bipartite graph.

The problem is to match teachers to classes.

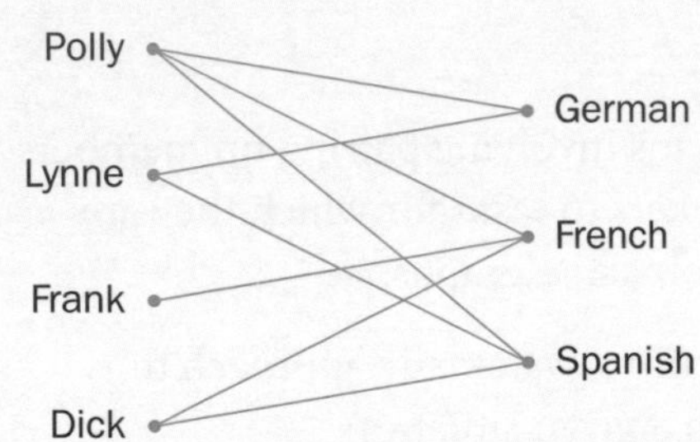

There are more teachers than classes, so one teacher will not be used. There are several ways the allocation could be made.

One possibility is:

Lynne teaches German
Frank teaches French
Dick teaches Spanish

This is shown on the graph.

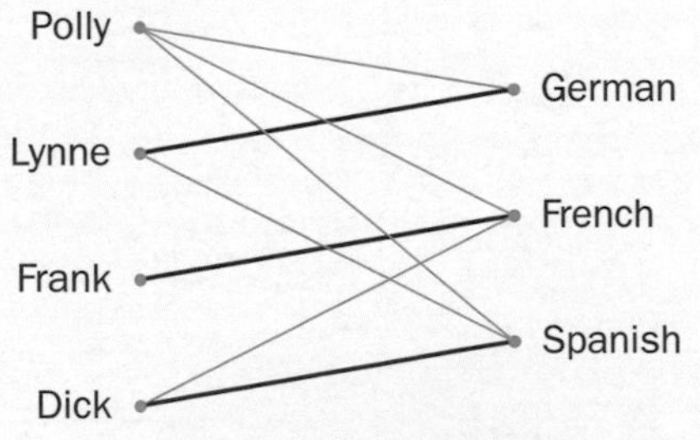

Have a go at finding other possible solutions.

For a bipartite graph, a **matching**, M, is a subset of the edges of the graph with the property that no two edges in M share a vertex.

You usually show a matching by drawing the edges in M using thicker, possibly coloured, lines.

Here are some examples of matchings for a particular graph.

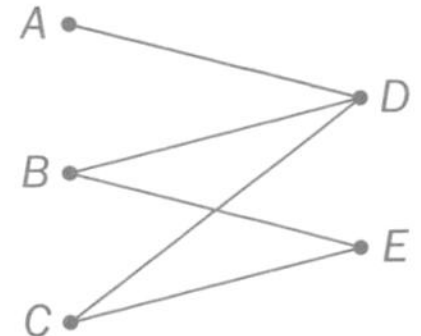

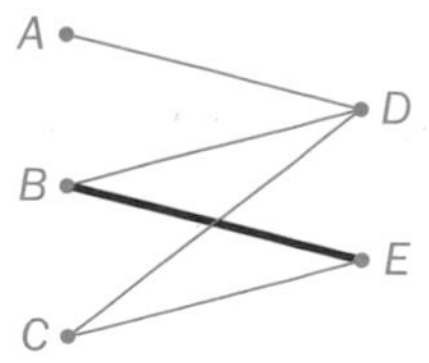

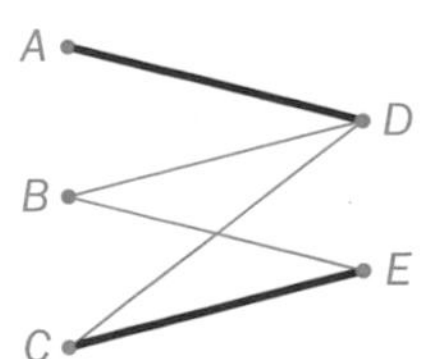

The matching 'pairs up' vertices in the two sets.

The first example is the trivial one in which M possesses no edges.

The third example shown is a maximal matching because both *D* and *E* have been matched. The matching could not have three edges because two of them would have to share a vertex.

In this example the maximal matching includes all of the vertices in the smaller set. This is **not** true for every maximal matching.

M is a maximal matching if it contains the greatest possible number of edges.

e.g. Consider the graph shown.

E and *F* are both connected only to *C*.
A matching cannot include both *CE* and *CF*, so one of these vertices must remain unconnected.

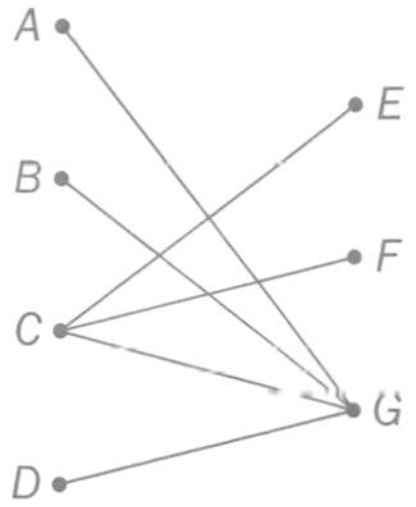

The maximal matching can only have two edges.
One possibility is shown.

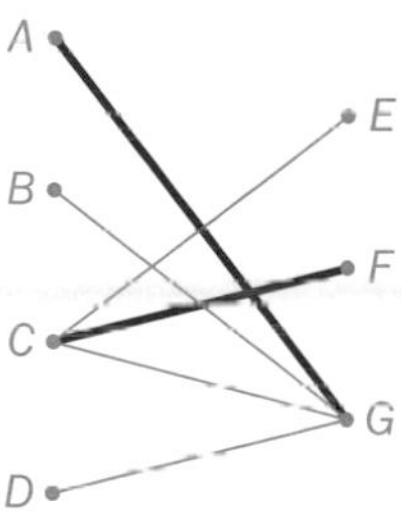

In many cases the two sets in the graph have the same number of vertices. You may then be able to find a complete matching.

e.g. This bipartite graph has equal sets of vertices.

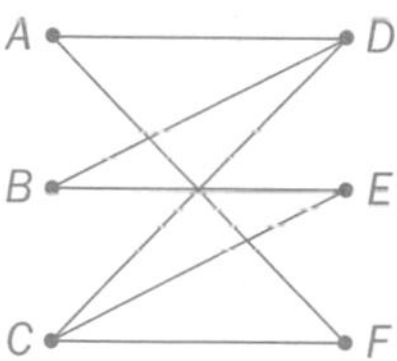

In this case it is possible to find a complete matching involving all six vertices.

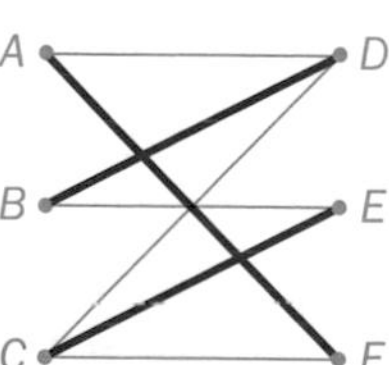

In a bipartite graph with equal numbers of vertices in each set, a matching, M, is a complete matching if M includes every vertex of the graph.

A complete matching between two sets of *n* vertices has *n* edges.

EXAMPLE 2

The table lists four workers – John, Kamlesh, Laura and Meera – and four tasks – *A*, *B*, *C* and *D* – and shows who is qualified for which task. Investigate whether it is possible to pair up each worker with a task for which they are qualified.

	A	*B*	*C*	*D*
J		✓		
K	✓	✓		
L	✓	✓	✓	✓
M	✓	✓		

Illustrate the problem using a bipartite graph.

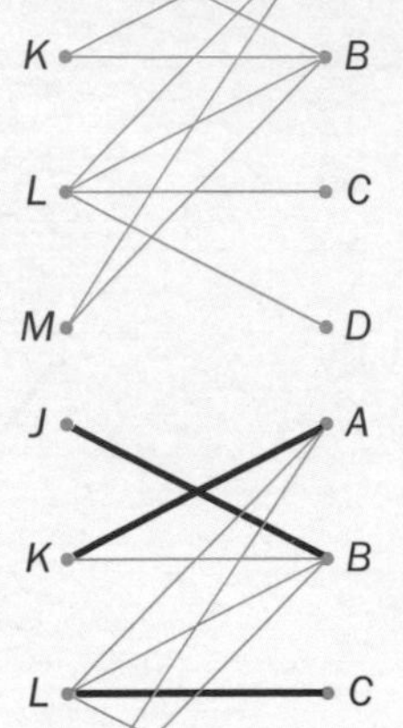

This highlights a difficulty. Both tasks *C* and *D* can only be done by Laura, so a complete matching is not possible.

The best that you can do is to match three people and three tasks. One way of doing this is illustrated.

Have a go at finding other possible solutions.

EXAMPLE 3

Show that there is a complete matching for the graph shown.

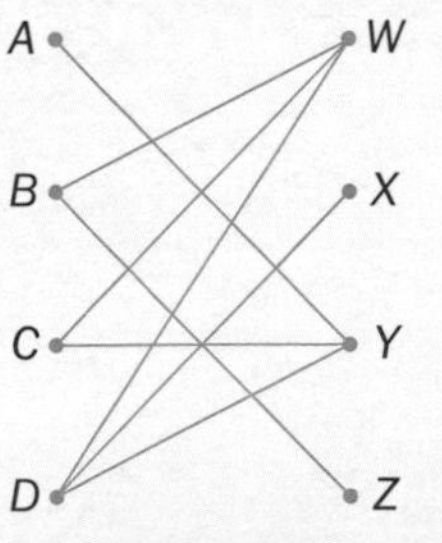

Vertices *A*, *X* and *Z* have only one connecting edge, so you are forced to choose edges *AY*, *BZ* and *DX*.
This means that the remaining pairing must be *CW*.

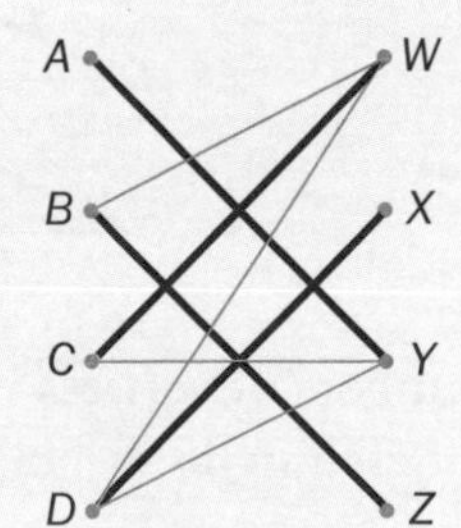

This is a complete matching.
It is the only one possible, as you had no choice at any stage.

Exercise 6.1

1 A school music teacher asks for pupils to put themselves forward to play in a group. The group will consist of four people – a drummer, a bassist, a guitarist and a vocalist. Five people come forward, and the parts they could take are shown in the table.

	Drums	Bass	Guitar	Vocals
Ali		✓		
Ben		✓	✓	
Cass	✓			✓
Dee		✓	✓	
Eve			✓	✓

a Show this information as a bipartite graph.

b Given that Ben and Dee are not prepared to play together, show that there are two possible line-ups.

2 Illustrate all possible maximal matchings for this bipartite graph.

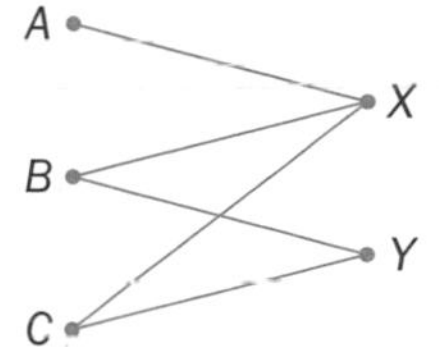

3 Find the two possible complete matchings for this bipartite graph.

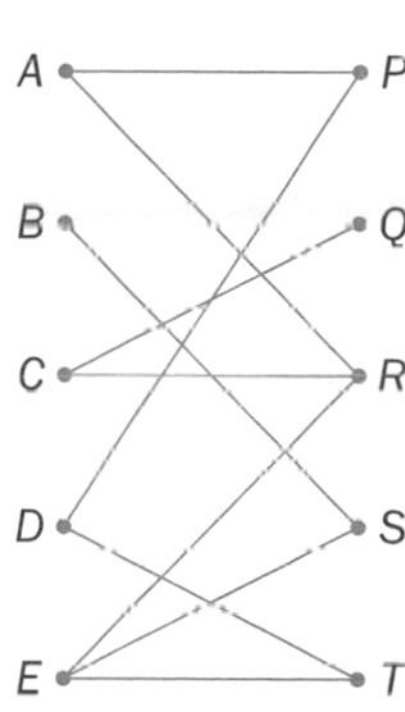

4 At a dog show there are five breed classes. Each class requires a judge. There are five judges available with knowledge of some of those breeds. Their expertise is summarised in the table.

Judge	Breeds
Mr Lee	Corgi, Beagle
Mrs Tweed	Samoyed, Corgi
Ms Pinkham	Beagle, Whippet
Mr Zapata	Corgi
Miss Floyd	Corgi, Whippet, Leonberger

a Show this information as a bipartite graph.

b Find a complete matching, explaining your working. Is it the only possible complete matching?

5 The owner of a row of four houses proposes to paint each of the front doors a different colour. She gives the tenants the chance to say which of the five possible colours they would definitely *not* want. This information is summarised in the table. A cross indicates that the tenant does not want that colour.

	House number			
	1	2	3	4
Blue		×	×	
Red	×			×
Green	×	×	×	
White		×		×
Yellow	×	×	×	

a Draw a bipartite graph to show which colours could go with which houses.

b Find a possible matching.

c Are there any colours which could not be used together?

6.2 Improving a matching

In complicated examples it can be quite difficult to see the maximal matching. You can easily find a partial matching. You need an algorithm which will

- improve on your matching if possible
- tell you if your matching is maximal.

Your matching consists of a set, M, of edges. You look for an **alternating path**.

Just picking one edge at random gives a partial matching, although you can usually do better than this.

> An **alternating path** joins an unmatched vertex in the left-hand set to an unmatched vertex in the right-hand set by edges which are alternately in and not in the set, M, of matched edges.

e.g. The diagram shows a bipartite graph with a partial matching. The set M = {*BP*, *CS*}.

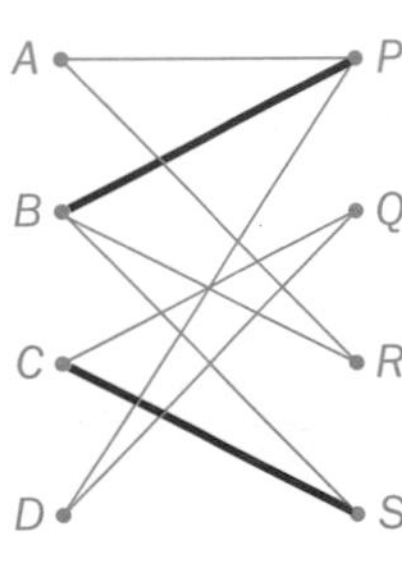

An example of an alternating path is

APBSCQ

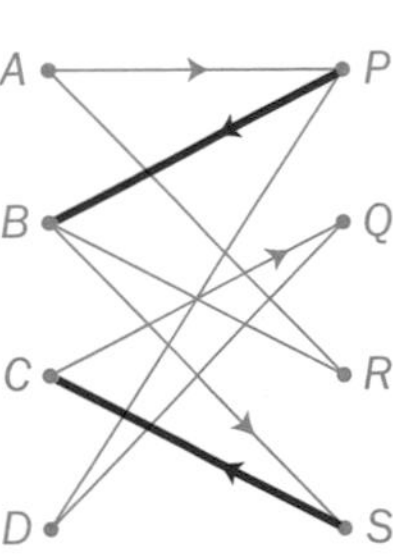

If you find an alternating path, you improve the matching by **changing the status** of the edges. The resulting matching contains one more edge than the previous matching.

> When you **change the status** of the edges of an alternating path, those edges belonging to M are removed from M, and those edges not in M are added to M.

In any alternating path, the number of edges in M is always one less than the number not in M. Changing the status will therefore increase the number of edges in the matching by one.

Edges shown in black are in M in this example.

e.g. The alternating path APBSCQ in the previous example

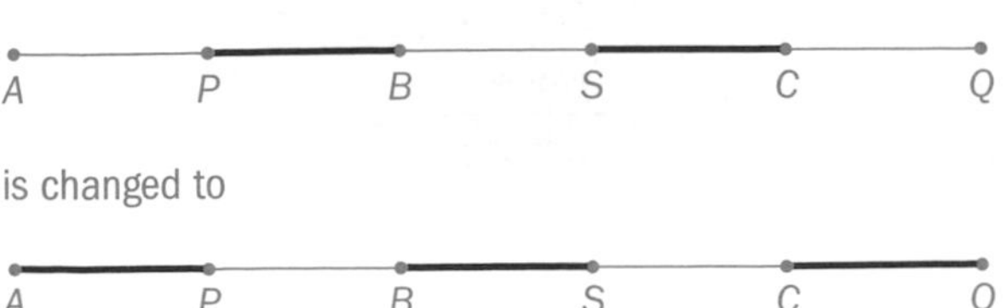

D1

The resulting matching is

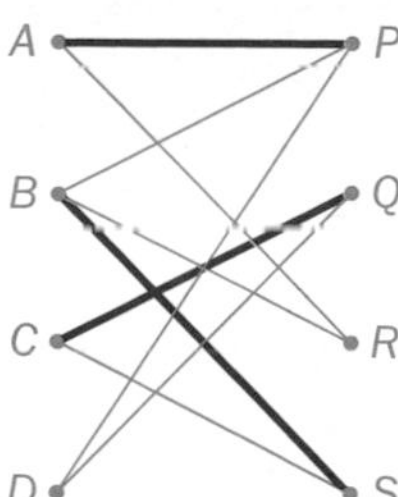

If you can find another alternating path, you can improve the matching further. If not, the matching is maximal.

e.g. Using the same example there is an alternating path *DPAR*.

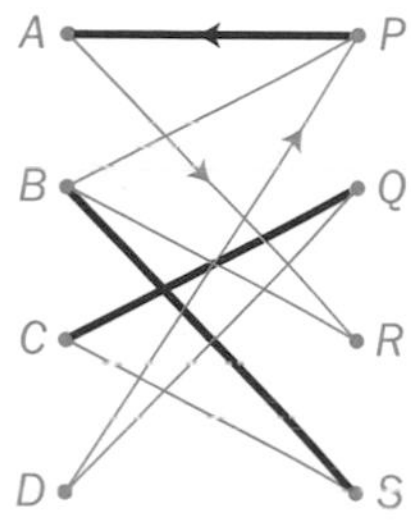

Changing the status gives

The resulting matching is now complete, because all of the vertices are connected.

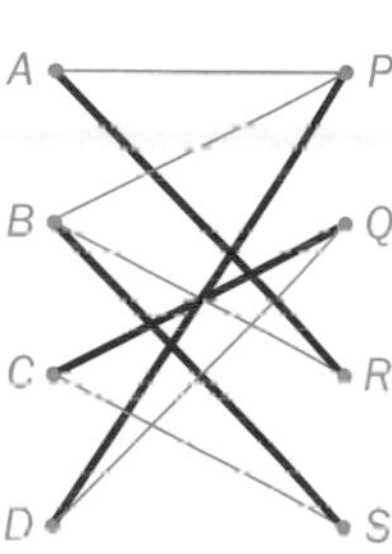

The process described above is the **maximum matching algorithm**.

Also called the **alternating path algorithm** or the **matching improvement algorithm**.

Maximum matching algorithm

Step 1 Find a matching M (the best you can easily find)

Step 2 Search for an alternating path.
If none exists, go to Step 4

Step 3 Form a new matching by changing the status of the edges in the alternating path, and go to Step 2

Step 4 Stop. The current matching is maximal.

If all vertices are connected, it is a complete matching.

Finding an alternating path

To find an alternating path you make a systematic search.

From each unconnected vertex in one set you explore all alternating paths. If you reach an unconnected vertex in the other set, you have made a **breakthrough**. If no starting vertex leads to a breakthrough, the matching is maximal.

There are other algorithms for finding an alternating path, but they lie outside the present syllabus.

A tree diagram can help in the search.

EXAMPLE 1

Search for an alternating path in the graph shown.

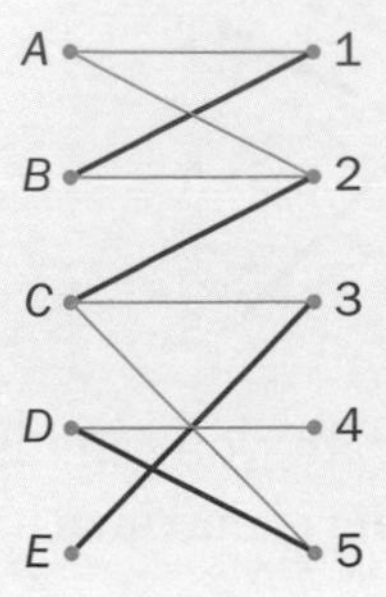

Start from vertex A.

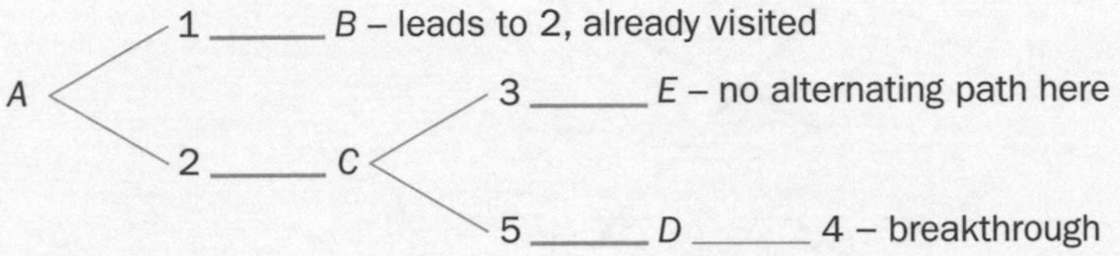

You have found an alternating path $A2C5D4$

D1

Exercise 6.2

1 Each of these diagrams shows a partial matching for a bipartite graph. Using the maximum matching algorithm, decide whether the given matching is maximal. If it is not, find a maximal matching.

a

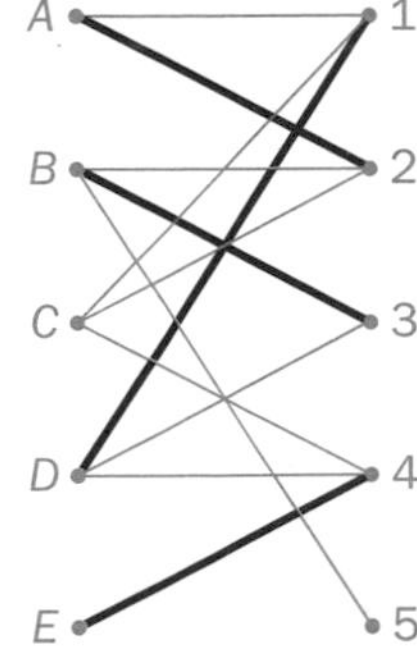

b

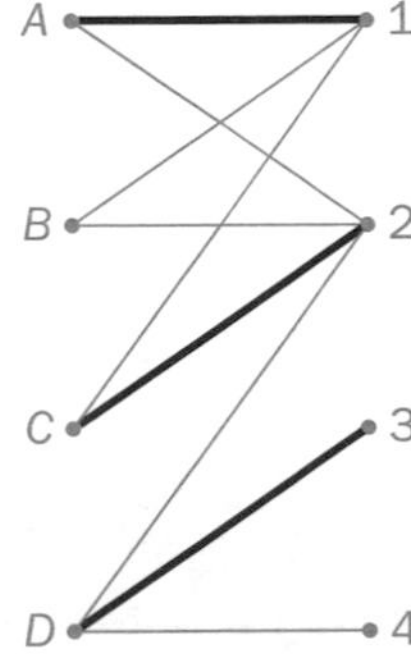

c

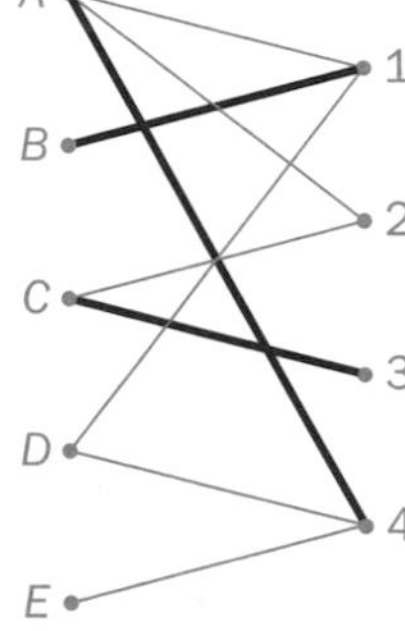

d

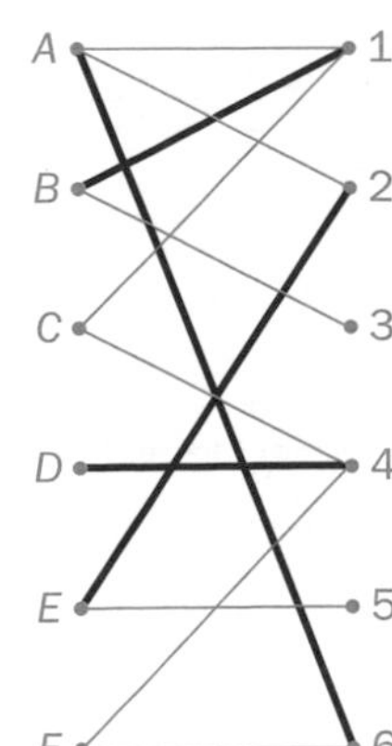

e

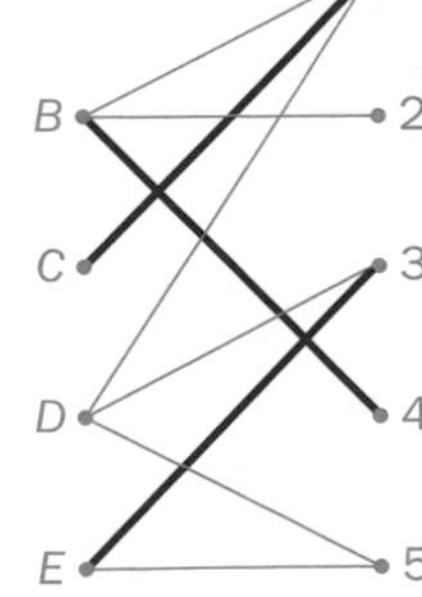

2 Four workers, A, B, C and D, are qualified for some of the tasks 1, 2, 3 and 4, as shown.

Worker	A	B	C	D
Task	1, 3	1, 2	2, 3, 4	2

The foreman starts by assigning B to task 1, C to task 3 and D to task 2, but then realises there is a problem. Draw a bipartite graph to model the situation and, using the maximum matching algorithm with the foreman's initial allocation as a starting point, find a complete matching of workers to tasks.

3 Aftab, Barry, Courtney and Dennis are to partner Edith, Fatima, Gladys and Hazel in a mixed doubles badminton tournament. Unfortunately they are all very choosy about who they play with. The table shows those pairings which will not cause problems.

	E	F	G	H
A	✓			✓
B	✓		✓	
C	✓			
D		✓	✓	

Aftab suggests 'I'll play with Edith, Barry can play with Gladys and Dennis with Fatima.' However, this leaves Courtney and Hazel, who don't get on.

Model the problem as a bipartite graph and, using the maximum matching algorithm with Aftab's suggestion as a starting point, find a complete matching.

4 Amy, Beulah, Caitlin and Daisy form a rowing team. There are four positions in the boat and the girls have the following preferences – Amy likes position 3 or 4, Beulah likes 1, 2 or 4, Caitlin likes 2 or 3 and Daisy likes 1 or 2.

a Show this information as a bipartite graph.

b Show on your graph the partial matching (Amy, 3), (Beulah, 1), (Daisy, 2).

c Show that there are two possible alternating paths starting from Caitlin. Hence find the two possible complete matchings for the graph.

5 A householder has a number of small building jobs that need doing, and gets quotes from four firms. The firms are not equipped to do all the jobs, and their quotes are shown in the table.

	Job 1	Job 2	Job 3	Job 4
Firm A	£200	–	£340	–
Firm B	£210	£110	–	–
Firm C	£180	–	–	£390
Firm D	–	£150	£320	£350

The jobs need to be done simultaneously, so it is necessary to give one job to each firm.

a Model the situation as a bipartite graph, and hence find all possible ways in which the jobs could be allocated.

b Hence decide which is the cheapest option for the householder.

D1

1 Amy, Bryn, Caspar, Delia and Everton are choosing between five available iced lollies – Lemon, Mango, Nectarine, Orange and Passion Fruit. To make it fair, they each state two flavours they would like, and try to arrange that everyone gets one of their choices. Their choices are shown in the table.

Child	Choices
Amy	Lemon, Nectarine
Bryn	Nectarine, Orange
Caspar	Mango, Orange
Delia	Lemon, Mango
Everton	Orange, Passion Fruit

a Show this information by means of a bipartite graph.

b Initially Bryn, Caspar, Delia and Everton are given the first of their two choices, but this leaves Amy disappointed. By using the maximum matching algorithm starting with this matching, find a complete matching of children to iced lollies.

2 Four people – John, Klaus, Lily and Mae – are in a dentist's waiting room. There are four magazine in the rack – *The Economist, Film Weekly, Gardener's World* and *Hello.* John will read *The Economist* or *Film Weekly.* Klaus likes *Film Weekly* or *Gardener's World.* Lily prefers *Film Weekly* or *Gardener's World.* Mae will read *The Economist* or *Hello.* Initially, John picks *Film Weekly,* Lily goes for *Gardener's World* and Mae has *The Economist.*

a Show this information on a bipartite graph.

b Show, by finding an alternating path, that the matching given can be improved to a complete matching.

3 **a** Draw the bipartite graph corresponding to the adjacency matrix shown.

	P	*Q*	*R*	*S*	*T*
A	0	1	0	0	0
B	0	1	0	0	0
C	1	1	0	1	0
D	0	0	1	0	1
E	0	0	0	1	1

b Starting from the initial matching *A*-*Q*, *C*-*S* and *D*-*T*, use the matching improvement algorithm to find a maximal matching. Explain how you know it is maximal.

c Is the maximal matching you found in part **b** a unique solution? Give a reason for your answer.

4 This bipartite graph shows a mapping between six people, Andy (*A*), David (*D*), Joan (*J*), Preety (*P*), Sally (*S*) and Trevor (*T*), and six tasks, 1, 2, 3, 4, 5 and 6.

The initial matching is *A* to 2, *D* to 1, *J* to 3 and *P* to 4.

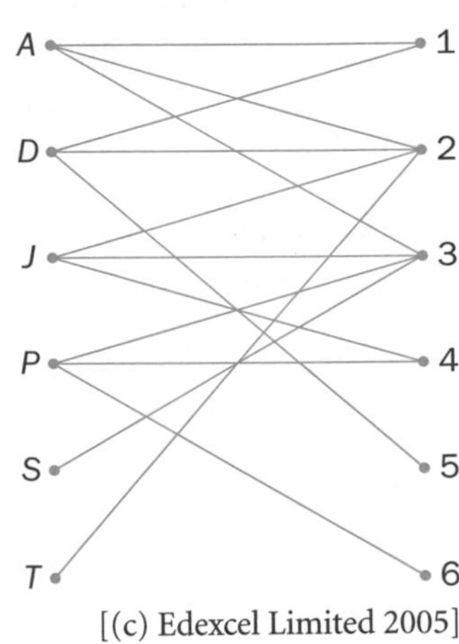

[(c) Edexcel Limited 2005]

a Indicate this initial matching in a distinctive way on a copy of the bipartite graph.

b Starting from this initial matching, use the maximum matching algorithm to find a complete matching. List clearly the alternating paths you use.

5 Ann, Bryn, Daljit, Gareth and Nickos have all joined a new committee. Each of them is to be allocated to one of five jobs 1, 2, 3, 4 or 5. The table shows each member's preferences for the jobs.

Ann	1 or 2
Bryn	3 or 1
Daljit	2 or 4
Gareth	5 or 3
Nickos	1 or 2

Initially Ann, Bryn, Daljit and Gareth are allocated the first job in their lists shown in the table.

a Draw a bipartite graph to model the preferences shown in the table and indicate, in a distinctive way, the initial allocation of jobs.

b Use the matching improvement algorithm to find a complete matching, showing clearly your alternating path.

c Find a second alternating path from the initial allocation. [(c) Edexcel Limited 2002]

6

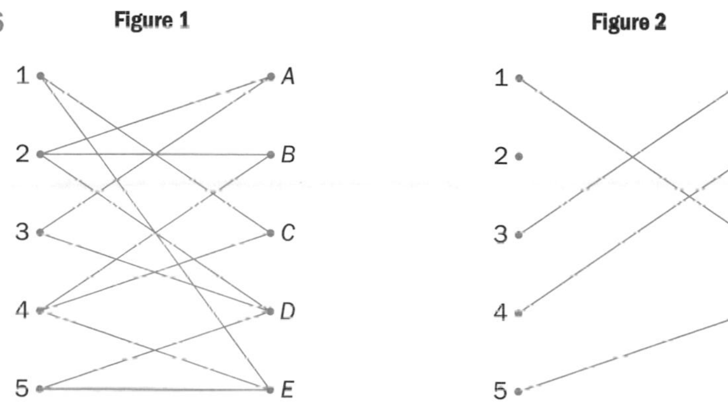

Five members of staff 1, 2, 3, 4 and 5 are to be matched to five jobs *A*, *B*, *C*, *D* and *E*. A bipartite graph showing the possible matchings is given in Fig. 1 and an initial matching *M* is given in Fig. 2.

There are several distinct alternating paths that can be generated from *M*. Two such paths are

2 —— *B* —— 4 —— *E*

and 2 —— *A* —— 3 —— *D* —— 5 —— *E*

a Use each of these two alternating paths, in turn, to write down the complete matchings they generate.

Using the maximum matching algorithm and the initial matching *M*

b find two further distinct alternating paths, making your reasoning clear. [(c) Edexcel Limited 2002]

6 Exit

Summary

Refer to

- You can use a bipartite graph to illustrate information regarding the possible connections between members of two sets.
  - A matching, M, is a subset of the edges of the graph such that no two edges of M share a vertex.
  - A matching is maximal if it contains the greatest possible number of edges.
  - A complete matching between sets of equal size is one in which every vertex is connected. 6.1
- For a given matching, M, an alternating path joins unmatched vertices in the two sets by a sequence of edges which alternately belong and do not belong to M. If no such path exists, the matching is maximal. 6.2
- The maximum matching algorithm improves on a given matching as follows.
  - Find an alternating path (•——•——•——•——•——•)
  - Change the status of the edges so that unmatched edges become matched, matched edges become unmatched (•——•——•——•——•——•). The resulting matching is an improvement. 6.2

Links

In recent years the need to find optimal solutions in areas such as traffic and transportation planning, logistics and scheduling of projects has increased dramatically.

As described in this chapter, simple matching problems arise in many resource allocation situations. For example, the techniques form the basis of some school timetabling software, matching classes to times to teachers to rooms etc.

In other situations a vertex may be matched to more than one other vertex, such as when a number of computer users log on to one of several servers in a distributed network. The speed of the servers will be affected by their workload, so the aim is to allocate them to achieve the most efficient service.

7 Linear programming

This chapter will show you how to

- express the problem of optimising a quantity such as profit or cost as a linear programming formulation
- solve two-variable problems graphically.

Introduction

Linear programming was developed in the 1940s to help with decision-making relating to finding the best possible value for some quantity which depends on a number of variables. The values the variables can take are subject to limitations.

The techniques are widely used in the commercial world, in particular in production planning, to obtain the maximum profit or to incur the minimum cost in a given situation.

Before you start

You should know how to:	Check in
1 Draw a straight-line graph.	1 Draw a graph showing the lines $y = 3x - 2$, $4x + 3y = 20$. Verify that they intersect at (2, 4).
2 Draw the graph of a linear inequality, shading the unwanted region.	2 Draw a graph to show the region satisfying the following inequalities. $x \geqslant 0$, $x + y \leqslant 12$, $y \geqslant 2x$
3 Solve linear simultaneous equations.	3 Solve the simultaneous equations $2x + 3y = 8$, $5x - y = 3$

7.1 Formulating the problem

Linear programming is about choosing the best combination of a number of quantities (variables) to achieve the best (optimal) outcome – typically to maximise profit or minimise cost.

In a real-life problem there are likely to be many variables. In this course you will not be required to deal with more than two.

First you translate the problem into mathematical terms – you produce a linear programming (LP) formulation.
To do this you

- identify the quantities you can vary.
  These are the decision variables.
- identify the limitations on the values of the decision variables.
  These are the constraints.
- identify the quantity to be optimised.
  This is the objective function.

Decision variables are sometimes called **control variables**. The problem is solved when you have found the value of these.

Constraints are usually inequalities involving the decision variables.

You write the objective function in terms of the decision variables.

Here are some examples of linear programming situations.

EXAMPLE 1

Cloggs Breakfast Cereals make two types of muesli – Standard and De Luxe.
1 kg of Standard contains 800 g of oat mix and 200 g of fruit mix. 1 kg of De Luxe has 600 g of oat mix and 400 g of fruit mix.
1 kg of Standard makes 60p profit.
1 kg of De Luxe makes 80p profit.

There are 3000 kg of oat mix and 1000 kg of fruit mix in stock. The company knows that they can sell at most 3500 kg of Standard and 2000 kg of De Luxe.
You need to decide how much of each type to make to maximise profit.
Write this problem as a linear programming formulation.

Summarise the information in a table:

	Oat mix (kg)	Fruit mix (kg)	Max sales (kg)	Profit (£ per kg)
1 kg Standard	0.8	0.2	3500	0.60
1 kg De Luxe	0.6	0.4	2000	0.80
Availability	3000	1000		

The **decision variables** are the amount, x kg, of Standard and the amount, y kg, of De Luxe.

The **constraints** come from the limitations on sales and the quantities of raw materials available.

Example 1 is continued on the next page.

EXAMPLE 1 (CONT.)

The upper limits on sales tell you that $x \leqslant 3500, y \leqslant 2000$

You will use $0.8x$ kg of oat mix for Standard muesli and $0.6y$ kg for De Luxe.
There is 3000 kg of oat mix, so $0.8x + 0.6y \leqslant 3000$

Simplify: $4x + 3y \leqslant 15\,000$

You will use $0.2x$ kg of fruit mix for Standard muesli and $0.4y$ kg for De Luxe.
There is 1000 kg of fruit mix, so $0.2x + 0.4y \leqslant 1000$

Simplify: $x + 2y \leqslant 5000$

There are two more, trivial, constraints, namely $x \geqslant 0, y \geqslant 0$

The **objective function** in this case is the profit, £P, which you need to maximise.

The profit on x kg of Standard and y kg of De Luxe is $P = 0.6x + 0.8y$

State the complete linear programming formulation:

Maximise $P = 0.6x + 0.8y$

Subject to
$$\begin{aligned} 4x + 3y &\leqslant 15\,000 \\ x + 2y &\leqslant 5000 \\ x &\leqslant 3500 \\ y &\leqslant 2000 \\ x \geqslant 0, y &\geqslant 0 \end{aligned}$$

EXAMPLE 2

A waste paper merchant has two processing plants.
Plant A can process 4 tonnes of waste paper and 1 tonne of cardboard per hour.
Plant B can process 5 tonnes of waste paper and 2 tonnes of cardboard per hour.
It costs £400 per hour to run plant A and £600 per hour to run plant B.
The merchant has agreed with the union that each plant gets at least one-third of the run time of every consignment.

A consignment arrives, consisting of 100 tonnes of waste paper and 35 tonnes of cardboard.
The problem is to share this between the two plants so that processing costs are minimised.
Write this as a linear programming formulation.

The solution to Example 2 is on the next page.

EXAMPLE 2 (CONT.)

Summarise the information in a table:

	Waste paper (tonnes)	Cardboard (tonnes)	Cost (£)
Plant A (1 hour)	4	1	400
Plant B (1 hour)	5	2	600
Availability	100	35	

The table is not necessary, but it can help to clarify your thinking.

The **decision variables** are the running times for the two plants.

Run A for x hours and B for y hours.

The **constraints** are the union agreement and the processing capacity of the plants.

The union agreement says $x \geqslant \frac{1}{3}(x+y)$ and $y \geqslant \frac{1}{3}(x+y)$

Simplify: $y \leqslant 2x$ and $x \leqslant 2y$

The two plants must process a total of at least 100 tonnes of paper in the time, giving

$4x + 5y \geqslant 100$

Similarly, they need to process at least 35 tonnes of cardboard, giving

$x + 2y \geqslant 35$

There are also the usual non-negativity constraints $x \geqslant 0$ and $y \geqslant 0$.

The **objective function** in this case is the cost, £C, which you need to minimise.

The cost of x hours of Plant A and y hours of Plant B is $C = 400x + 600y$

State the complete linear programming formulation:

Minimise $C = 400x + 600y$

Subject to
$$\begin{aligned} y &\leqslant 2x \\ x &\leqslant 2y \\ 4x + 5y &\geqslant 100 \\ x + 2y &\geqslant 35 \\ x \geqslant 0, y &\geqslant 0 \end{aligned}$$

The next example includes a different sort of constraint.

D1

EXAMPLE 3

A manufacturer makes three types of dining chair.
All chairs pass through three workshops for cutting, assembly and finishing.
Each workshop runs for a working week of 40 hours.
The times in the workshops for each type of chair and the profit gained from each are shown in the table.

Chair type	Cutting (hours)	Assembly (hours)	Finishing (hours)	Profit (£)
A	0.5	1	0.75	60
B	0.75	1.25	0.75	80
C	0.5	0.75	0.5	50

The aim is to maximise the weekly profit. Write this as a linear programming formulation.

The **decision variables** are numbers of each chair type.
Make x of Type A, y of Type B and z of Type C.

The **constraints** are the times available in each workshop.

The time in the cutting shop means that

$$0.5x + 0.75y + 0.5z \leqslant 40$$

Simplify: $$2x + 3y + 2z \leqslant 160$$

The time in the assembly shop means that

$$x + 1.25y + 0.75z \leqslant 40$$

Simplify: $$4x + 5y + 3z \leqslant 160$$

The time in the finishing shop means that

$$0.75x + 0.75y + 0.5z \leqslant 40$$

Simplify: $$3x + 3y + 2z \leqslant 160$$

There are the usual non-negativity constraints $x \geqslant 0$, $y \geqslant 0$ and $z \geqslant 0$.

In this example there is also the constraint that you can't sell fractions of a chair, so x, y and z are integers.

The **objective function** is the profit, £P, which you need to maximise.
The profit is $P = 60x + 80y + 50z$

State the complete linear programming formulation:

Maximise $P = 60x + 80y + 50z$

Subject to
$$2x + 3y + 2z \leqslant 160$$
$$4x + 5y + 3z \leqslant 160$$
$$3x + 3y + 2z \leqslant 160$$
$$x \geqslant 0,\ y \geqslant 0,\ z \geqslant 0$$
x, y and z are integers.

D1

Example 4 shows how you can use linear programming to solve blending problems.

EXAMPLE 4

A company supplying vegetable oil buys from two sources, A and B. The oils are already a blend of olive oil, sunflower oil and other vegetable oils. The table shows the proportions, price and minimum weekly order of these.

	Olive oil	Sunflower oil	Other	Cost (p per litre)	Minimum order (litres)
A	50%	10%	40%	25	35 000
B	20%	60%	20%	20	50 000

The company wants to make a blend with at least 30% olive oil and at least 40% sunflower oil. They want to produce at least 90 000 litres per week, and to minimise the cost.
Write this as a linear programming formulation.

The **decision variables** are the amounts of the two types to use.
Use x litres of A and y litres of B.
The **constraints** are the proportions of olive and sunflower oil required, the minimum order quantities and the total production required.
The amount of olive oil is $(0.5x + 0.2y)$ in a total production of $(x + y)$.

The need for 30% olive oil gives $\frac{0.5x + 0.2y}{x + y} \geqslant 0.3$

Simplify: $y \leqslant 2x$

The amount of sunflower oil is $(0.1x + 0.6y)$ in a total production of $(x + y)$.

The need for 40% sunflower oil gives $\frac{0.1x + 0.6y}{x + y} \geqslant 0.4$

Simplify: $2y \geqslant 3x$

The minimum order requirements give $x \geqslant 35\,000$ and $y \geqslant 50\,000$

The total production constraint is $x + y \geqslant 90\,000$

The **objective function** is the cost, £C, which you need to minimise.
The cost is $C = 25x + 20y$

State the complete linear programming formulation:

Minimise $C = 25x + 20y$

Subject to

$$\begin{aligned} y &\leqslant 2x \\ 2y &\geqslant 3x \\ x &\geqslant 35\,000 \\ y &\geqslant 50\,000 \\ x + y &\geqslant 90\,000 \end{aligned}$$

D1

Exercise 7.1

For each of these situations state the problem as a linear programming formulation.

1 Roger Teeth Ltd make fruit drinks of two types, Econofruit and Healthifruit, consisting of fruit juice, sugar syrup and water. The proportions of fruit juice and sugar syrup in the two drinks are shown in the table.

	Fruit juice	Sugar syrup
Econofruit	20%	50%
Healthifruit	40%	30%

There are 20 000 litres of fruit juice and 30 000 litres of sugar syrup in stock (and unlimited water). The profit per litre is 30p for Econofruit and 40p for Healthifruit. The company wishes to maximise profit.

2 A club, with 80 members, is organising a trip. The intention is to hire vehicles they can drive themselves and travel in convoy. Only eight of the members are prepared to drive. A car, which can carry five people including the driver, costs £20 per day to hire. A minibus, which can carry 12 people including the driver, costs £60 per day to hire. The club wishes to minimise the hire costs.

3 A farmer has 75 hectares of land on which to grow a mixture of wheat and potatoes. The costs and profits involved are shown in the table.

	Labour (man-hours per ha)	Fertiliser (kg per ha)	Profit (£ per ha)
Wheat	30	700	80
Potatoes	50	400	100

There are 2800 man-hours of labour and 40 tonnes of fertiliser available. The aim is to maximise profit.

4 A fruiterer is making up baskets of fruit, which are advertised as containing a mixture of oranges, apples and pears with at least 30 fruit in all. There must be at least as many apples as oranges, and at least twice as many apples as pears. Oranges cost 20p each, apples 12p each and pears 15p each. The aim is to minimise the cost.

5 Whisky mac is a mixture of whisky and ginger wine. Whisky is 40% alcohol and costs £12 per litre. Ginger wine is 12% alcohol and costs £5 per litre. A barkeeper wishes to minimise the cost of making a whisky mac, which must be at least 100 ml of liquid, at least 20% alcohol and contain at most 30 ml of alcohol.

6 Petmax Petfoods make two types of tinned cat food. 'Vitapuss' contains 40% fish meal and 60% meat. 'Pamper' has 30% fish meal and 70% meat. Both brands sell in 500 g tins, Vitapuss for 60p and Pamper for 80p. They buy the fish meal at 15p per kg and the meat at 20p per kg. At the start of a production run they have 4000 kg of fish meal and 6500 kg of meat in stock, and they wish to know how much of each brand to produce to maximise their profit from the day's production.

7 A trader is buying goods from a warehouse and taking them by van to his shop. He wants shampoo and cleaning powder. These are packed in cases of the same size. A case of shampoo weighs 6 kg and he can make a profit of £20 per case. A case of cleaning powder weighs 4 kg and his profit will be £14 per case. His van has room for 60 cases and can carry a maximum load of 280 kg. He needs to know how much of each commodity to buy in order to maximise his profit.

8 An investor has up to £20 000 to invest. She can buy 'safe' bonds yielding 5% interest or 'risky' shares yielding 10% interest. Both types of shares cost £1 each. She wants to make at least 8% interest overall, and as she is a cautious investor she wants her investment in bonds to be as great as possible.

Express her investment as a linear programming formulation.

9 Pete Potts Horticultural Supplies sell two grades of grass seed, each a mixture of perennial ryegrass (PR) and creeping red fescue (CRF). Regular Lawn Mix is 70% PR and 30% CRF. Luxury Lawn Mix is equal proportions of each.
Pete buys PR at £4 per kg and CRF at £5 per kg. He sells Regular mix at £6 per kg and Luxury mix at £7 per kg.
He has 8000 kg of PR and 6000 kg of CRF in stock.

Express the problem of maximising his profit as a linear programming formulation.

10 My electric bicycle has two settings. On setting *A* I have to pedal, with the motor providing some assistance. On setting *B* the motor does all the work. Setting *A* gives a speed of 4 m s^{-1} and uses 6 joules of battery energy per metre. Setting *B* gives a speed of 6 m s^{-1} and uses 9 joules per metre. The battery can store 45 000 joules. I wish to travel as far as possible in 20 minutes.

I travel x metres on setting *A* and y metres on setting *B*.

Express the problem as linear programming formulation, simplifying the inequalities to have integer coefficients.

11 Floors 'R' Us Ltd sells three different types of carpet: Floral, Modern, and Swirl. The different patterns are all made using green, red and blue dye.

For each roll of carpet:
Floral requires 2 units of green, 3 units of red and 1 unit of blue
Modern requires 1 unit of green, 2 units of red and 2 units of blue
Swirl requires 2 units of green, 2 units of red and 2 units of blue.

There are up to 40 units of green, 60 units of red and 30 units of blue available every day.
Profits of £60, £40 and £50 are made on each roll of Floral, Modern and Swirl respectively.

Express the problem of maximising the company's profit as a linear programming formulation.

D1

7.2 Solving problems graphically

Problems with two decision variables can be solved graphically.

e.g. Consider the linear programming formulation for the muesli production problem in Example 1 (page 108):

Maximise $P = 0.6x + 0.8y$

subject to the constraints

$$x \leqslant 3500$$
$$y \leqslant 2000$$
$$4x + 3y \leqslant 15\,000$$
$$x + 2y \leqslant 5000$$
$$x \geqslant 0, y \geqslant 0$$

where x kg of Standard and y kg of De Luxe muesli were produced.

Draw a graph illustrating the constraints:
First $x \leqslant 3500$ and $y \leqslant 2000$.

Draw the lines $x = 3500$, $y = 2000$.
Shade the regions which are **not** needed.

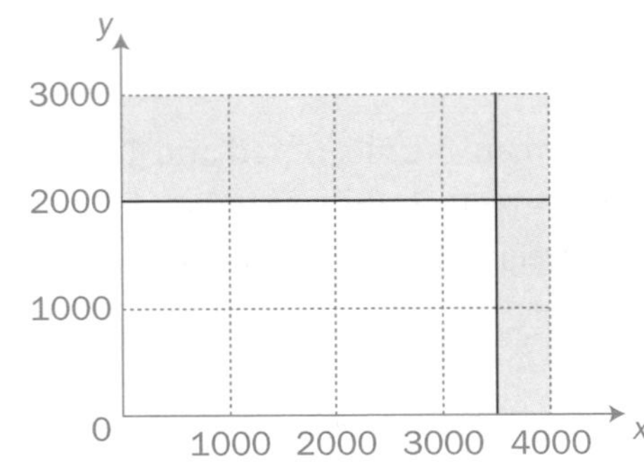

Next put in $4x + 3y \leqslant 15\,000$.
Draw the line $4x + 3y = 15\,000$.
Shade the region which does not satisfy the inequality.

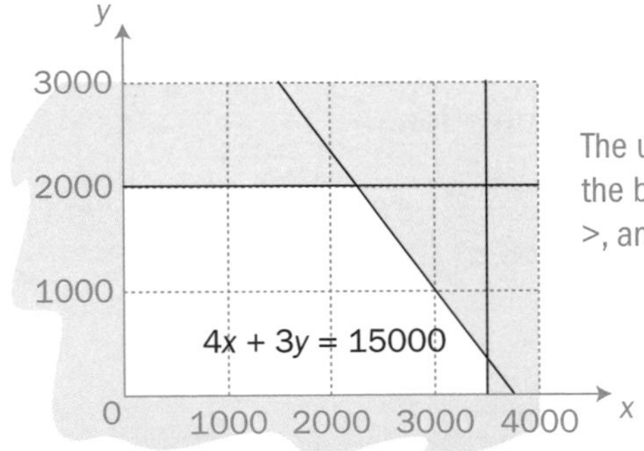

The usual convention is to draw the boundary line dotted for < or >, and continuous for $\leqslant$ and $\geqslant$.

Finally add $x + 2y \leqslant 5000$, $x \geqslant 0$, $y \geqslant 0$.

All allowable combinations of x and y lie in the unshaded region of the graph (including its boundary lines).
This is called the **feasible region**.

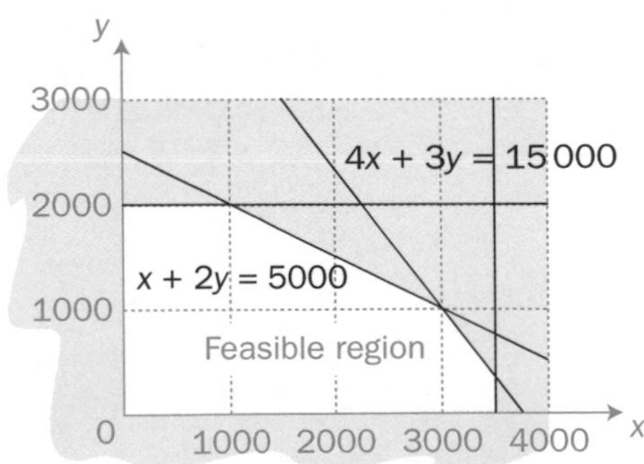

The feasible region is the set of (x, y) values which satisfy all the constraints.
It is the unshaded region on the graph.

D1

You can now illustrate the objective function $P = 0.6x + 0.8y$

Suppose you chose the production plan $x = 1000$, $y = 1000$.

This would give

$P = 0.6 \times 1000 + 0.8 \times 1000 = 1400$

This plan gives £1400 profit.

This is not a very good production plan, as lots of oat mix and fruit mix is left over.

There are other production plans giving $P = 1400$, for example $x = 0$, $y = 1750$ or $x = 2000$, $y = 250$.

These plans all lie on the line

$$0.6x + 0.8y = 1400$$

Draw this line on your graph:

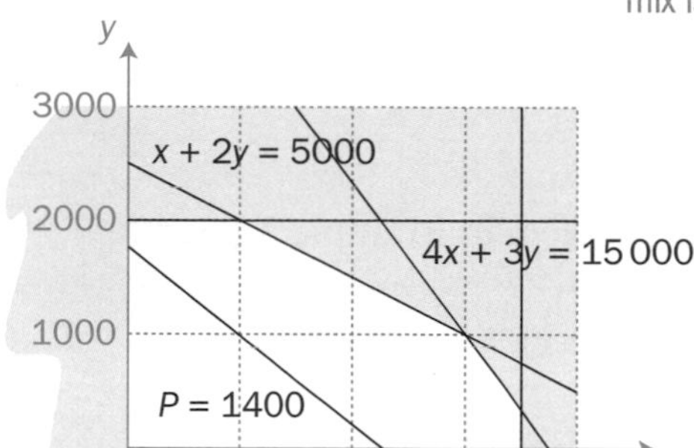

Similarly, you could draw the line

$$0.6x + 0.8y = 1800$$

showing all production plans giving a profit of £1800.

You have drawn two possible positions of the **objective line**.

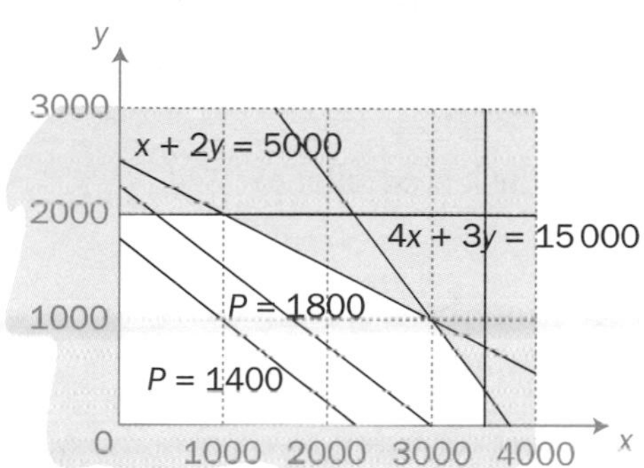

The objective line is a line joining all points (x, y) for which the objective function takes a specified value.

Your aim is to find the position of the objective line corresponding to the greatest possible profit.

Profit P increases as the objective line moves to the right (always keeping the same gradient). As long it crosses the feasible region, there is a production plan which will give that profit.

You need to find the extreme position of the line which still includes a point of the feasible region.

As the objective line is moved to the right, its last contact with the feasible region will be at a vertex.

If the objective line is parallel to one of the boundaries, then all points on that boundary will give the optimal value for the objective function.

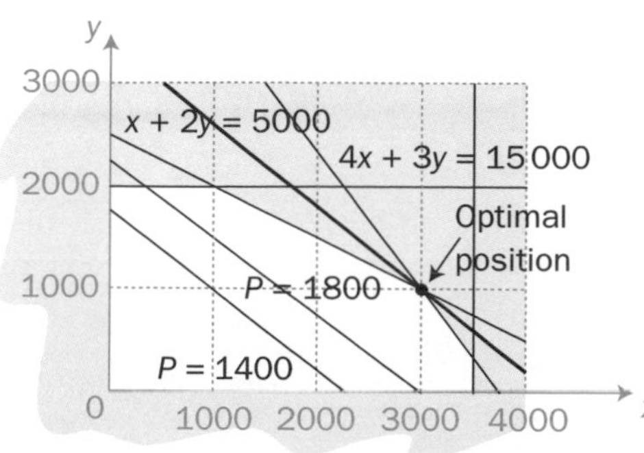

Slide the objective line to the right.

It leaves the feasible region where the lines $x + 2y = 5000$ and $4x + 3y = 15\,000$ intersect.

Solving simultaneous equations you get the point (3000, 1000).

This gives

$P = 0.6 \times 3000 + 0.8 \times 1000 = 2600$

If the choice of vertex is not obvious, you find the coordinates of the likely vertices and test which one gives the best value of the objective function.

The best production plan is 3000 kg of Standard muesli and 1000 kg of De Luxe muesli, giving a profit of £2600.

D1

EXAMPLE 1

Minimise $C = 400x + 600y$

Subject to

$$x \leqslant 2y$$
$$y \leqslant 2x$$
$$4x + 5y \geqslant 100$$
$$x + 2y \geqslant 35$$
$$x \geqslant 0, y \geqslant 0$$

(This is the waste paper problem in Example 2, page 109).

Draw a graph showing the constraints, with the unwanted regions shaded:

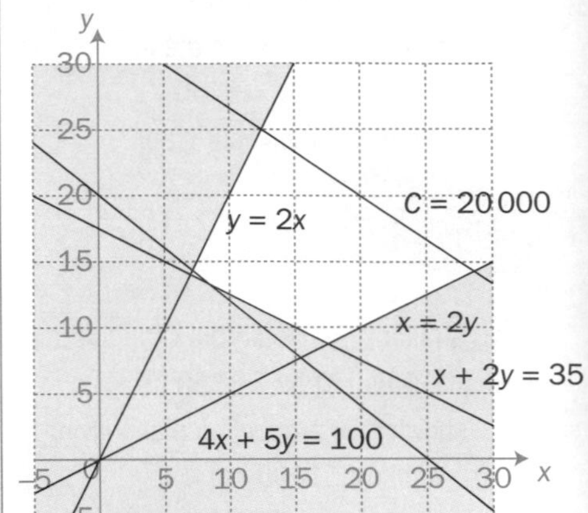

Draw a possible position of the objective line.
For example, the point (20, 20) is in the feasible region and gives $C = 20\,000$.
Draw the line $400x + 600y = 20\,000$.

Moving the objective line to the left parallel to itself reduces the value of C.

In this case it is hard to see from the graph which of the three vertices gives the minimum value of C.

Find the coordinates of the vertices using simultaneous equations.

Solving $y = 2x$ and $4x + 5y = 100$ gives $x = 7\frac{1}{7}, y = 14\frac{2}{7}$

This gives $C = 400 \times 7\frac{1}{7} + 600 \times 14\frac{2}{7} = 11\,428\frac{4}{7}$

Solving $4x + 5y = 100$ and $x + 2y = 35$ gives $x = 8\frac{1}{3}, y = 13\frac{1}{3}$

This gives $C = 400 \times 8\frac{1}{3} + 600 \times 13\frac{1}{3} = 11\,333\frac{1}{3}$

Solving $x + 2y = 35$ and $x = 2y$ gives $x = 17\frac{1}{2}, y = 8\frac{3}{4}$

This gives $C = 400 \times 17\frac{1}{2} + 600 \times 8\frac{3}{4} = 12\,250$

The minimum value of C is $11\,333\frac{1}{3}$, when $x = 8\frac{1}{3}, y = 13\frac{1}{3}$.

The merchant should run plant A for $8\frac{1}{3}$ hours and plant B for $13\frac{1}{3}$ hours.

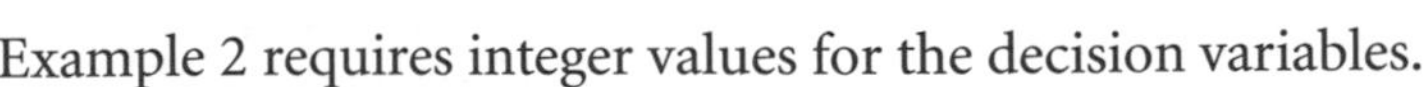
Example 2 requires integer values for the decision variables.

D1

EXAMPLE 2

Maximise $P = 42x + 20y$
Subject to $x + 2y \leqslant 14$
$3x + y \leqslant 20$
$x \geqslant 0, y \geqslant 0$
x and y are integers.

Draw a graph showing the constraints, shading the unwanted region:

Make sure you include all of the constraints.

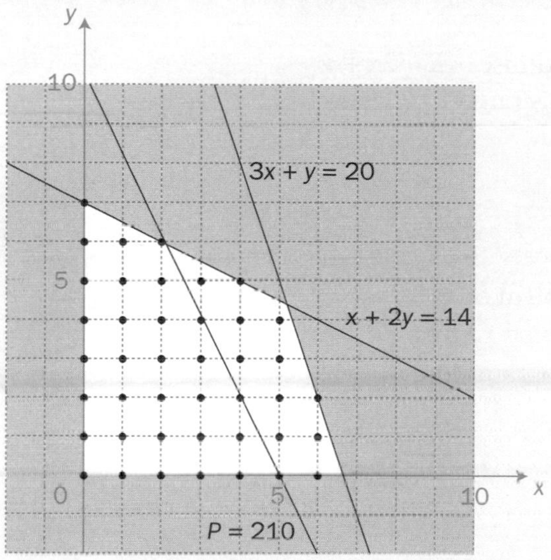

Indicate by dots the integer points (x, y) in the feasible region.

Draw one possible position of the objective line, for example $P = 42x + 20y = 210$.

As you move the objective line to the right, it leaves the feasible region at the intersection of the lines $x + 2y = 14$ and $3x + y = 20$.
But these lines intersect at (5.2, 4.4), which is not an integer point.

You need to test all likely integer points near (5.2, 4.4) to find which gives the maximum value of P.

These three points are the ones closest to or on the boundaries of the region.

The obvious points to test are (4, 5), (5, 4) and (6, 2).

$x = 4$ and $y = 5$ gives $P = 42 \times 4 + 20 \times 5 = 268$
$x = 5$ and $y = 4$ gives $P = 42 \times 5 + 20 \times 4 = 290$
$x = 6$ and $y = 2$ gives $P = 42 \times 6 + 20 \times 2 = 292$

Hence the optimum value of P is 292, obtained when $x = 6$ and $y = 2$.

D1

Exercise 7.2

1 For each of these linear programming problems, draw a graph showing the feasible region, and one possible position of the objective line. Hence find the optimal value of the objective function and the corresponding values of x and y.

a Maximise $P = 3x + 2y$
subject to
$3x + 4y \leqslant 120$
$3x + y \leqslant 75$
$x \geqslant 10, y \geqslant 5$

b Maximise $P = 4x + y$
subject to
$x + y \leqslant 10$
$2x + y \leqslant 16$
$x \geqslant 0, y \geqslant 0$

c Minimise $C = 5x + 4y$
subject to
$15x + 8y \geqslant 90$
$y \geqslant x$
$y \leqslant 2x$
x, y are integers

d Maximise $R = 2x + y$
subject to
$x + y \leqslant 14$
$x + 2y \leqslant 20$
$2x + 3y \leqslant 32$
$x \geqslant 0, y \geqslant 0$

2 In Example 4 on page 112 you explored the problem of a company blending oil from two sources. It obtained x litres from source A and y litres from B. The linear programming formulation was

Minimise $C = 25x + 20y$
subject to
$y \leqslant 2x$
$2y \geqslant 3x$
$x \geqslant 35\,000$
$y \geqslant 50\,000$
$x + y \geqslant 90\,000$

Use graphical methods to decide how much oil the company should buy from each source.

3 In Exercise 7.1 question **1** you expressed as a linear programming (LP) formulation the problem of deciding the most profitable quantities of Econofruit and Healthifruit drinks to make. Draw a graph showing the feasible region and hence find the optimum combination.

4 In Exercise 7.1 question **2** you expressed as an LP formulation the problem of deciding the cheapest combination of minibuses and cars to hire. Draw a graph showing the feasible region and hence find the optimum combination.

5 In Exercise 7.1 question **3** you expressed as an LP formulation the problem of maximising profit from growing wheat and potatoes. Use graphical methods to obtain the optimum planting strategy.

6 A building firm has a plot of land with area 6000 m^2. The intention is to build a mixture of houses and bungalows. A house occupies 210 m^2 and a bungalow occupies 270 m^2. Local planning regulations limit the total number of dwellings that can be built to 25, and insist that there must be no more than 15 of either type. Each house will realise £20 000 profit, and each bungalow £25 000. Formulate this situation as a linear programming problem and find graphically the best combination of dwellings.

7 It is planned to take 50 people on a trip. The party comprises x senior staff, y trainees and z children. There must be at least one adult to every two children, and at least one senior staff member to every trainee. There must be at least five trainees and at least ten senior staff. The cost of the trip is £20 for each senior staff member, £15 for each trainee and £12 for each child. It is required to minimise the total cost.

a Express this as an LP formulation in x, y and z.

b Using the fact that $x + y + z = 50$, show that your formulation can be expressed as

$$\begin{aligned} &\text{Minimise} && C = 8x + 3y + 600 \\ &\text{subject to} && 3x + 3y \geqslant 50 \\ &&& x \geqslant y \\ &&& x \geqslant 10,\ y \geqslant 5 \\ &&& x, y \text{ are integers} \end{aligned}$$

c Use graphical methods to find the optimum combination of people.

8 A dog food manufacturer makes three types of chew, each 10 g in weight, which are made from different proportions of two basic ingredients. The table shows this together with the amount of the two ingredients in stock, and costs for the three types of chew.

	Ingredient 1	Ingredient 2	Cost (p per chew)
Chew A	8 g	2 g	1.8
Chew B	6 g	4 g	1.6
Chew C	5 g	5 g	1.5
Availability	800 kg	400 kg	

The manufacturer wants to make 1000 packets of mixed chews. Each must contain 60 chews, and there must be no more than 30 of each type of chew in a packet. Find the best combination of chew types in each packet to minimise cost.

Review 7

1 A manufacturer makes two brands of blackcurrant jam – 'Value' and 'Luxury'. It takes 0.4 kg of blackcurrants and 0.8 kg of sugar to make 1 kg of Value jam, while 1 kg of Luxury jam requires 0.6 kg of blackcurrants and 0.6 kg of sugar. The profit is 10p per kg on Value jam and 12p per kg on Luxury jam. There are 480 kg of blackcurrants and 810 kg of sugar available. The aim is to maximise the profit.

a Express this situation as a linear programming formulation. Simplify the inequalities to give integer coefficients.

b Draw a graph to show the constraints, indicating clearly the feasible region. Draw the objective line corresponding to a profit of £60.

c Find the maximum profit, and the corresponding quantities of the two brands of jam.

2 The Bonzo Manufacturing Company makes model cars and lorries. Each car sells at a profit of £2.50 and each lorry sells at a profit of £3.00. Three departments, Manufacturing (Dept *A*), Assembly (Dept *B*) and Finishing (Dept *C*), are involved in the production of the models. The times, in hours, that the cars and lorries are in each department are shown in the table.

	Car	Lorry
Dept *A*	1.50	3.00
Dept *B*	2.00	1.00
Dept *C*	0.25	0.25

In a given week, 45 hours are available in Department *A*, 35 hours in Department *B* and 5 hours in Department *C*. The manufacturer wishes to maximise his profit £P in this week.

Let x be the number of cars made, and y be the number of lorries made. You may assume that all models made can be sold.

a Model this situation as a linear programming problem, giving each inequality in its simplest form with integer coefficients.

b Display the inequalities on a graph and identify the feasible region.

c By testing each vertex of the feasible region, obtain the maximum profit and the corresponding values of x and y.

d State which department has unused time and calculate this time.

[(c) Edexcel Limited 2003]

3 Becky's bird food company makes two types of bird food. One type is for bird feeders and the other for bird tables. Let x represent the quantity of food made for bird feeders and y represent the quantity of food made for bird tables. Due to restrictions in the production process, and known demand, the following constraints apply.

$$x + y \leqslant 12$$
$$y \leqslant 2x$$
$$2y \geqslant 7$$
$$y + 3x \geqslant 15$$

a Draw a graph to show these constraints and label the feasible region R.

The objective is to minimise $C = 2x + 5y$.

b Solve this problem, making your method clear. Give, as fractions, the value of C and the amount of each type of food that should be produced.

Another objective (for the same constraints given above) is to maximise $P = 3x + 2y$, where the variables must take integer values.

c Solve this problem, making your method clear. State the value of P and the amount of each type of food that should be produced. [(c) Edexcel Limited 2004]

D1

4 The Young Enterprise Company 'Decide' is going to produce badges to sell to decision maths students. It will produce two types of badge.

Badge 1 reads 'I made the decision to do maths' and Badge 2 reads 'Maths is the right decision'.

'Decide' must produce at least 200 badges and has enough material for 500 badges. Market research suggests that the number produced of Badge 1 should be between 20% and 40% of the total number of badges made.

The company makes a profit of 30p on each Badge 1 sold and 40p on each Badge 2. It will sell all that it produces and wishes to maximise its profit.

Let x be the number produced of Badge 1 and y be the number of Badge 2.

a Formulate this situation as a linear programming problem, simplifying your inequalities so that all the coefficients are integers.

b Draw a graph showing the feasible region.

c Using your graph, advise the company on the number of each badge it should produce. State the maximum profit 'Decide' will make. [(c) Edexcel Limited 2004]

7 Exit

Summary

Refer to

- You use linear programming to optimise an outcome which depends on the values of a number of variables. You are usually required to maximise a profit or minimise a cost. 7.1
- Problems with two decision variables x and y can be solved graphically. 7.2

Links

Manufacturing companies use linear programming to ensure maximum profits while satisfying their customers.

A confectionery company producing boxes of assorted chocolates needs to consider the weight, production cost and customer rating of each type of chocolate. Standard constraints would then be a minimum number of each type in each box, the total weight of the contents of the box and a maximum limit of the cheaper varieties (to ensure a reasonable mixture), etc. The company could then try out different approaches such as minimising the cost subject to meeting a minimum constraint on customer rating, or maximising customer rating subject to a minimum cost.

8 Critical path analysis

This chapter will show you how to

- model the activities involved in a project by means of a network
- calculate the minimum completion time for a project
- identify the critical activities – those activities which must not be delayed if the project is to be completed on time
- illustrate the project timetable using a Gantt chart
- schedule the project by allocating activities to workers.

Introduction

You can break down any large project into separate activities which are inter-related.
Some of the activities cannot start until others are completed.
Some of the activities are critical, in that any delay has a knock-on effect on the overall timetable of the project.

This chapter looks at the techniques for analysing projects – known as Critical Path Analysis (CPA) – which were developed in the 1950s, and have since become commonplace in a wide range of situations, from large construction projects to maintenance schedules for aircraft.

8.1 Modelling the problem

When devising the timetable for a project, you need

- a list of the activities involved
- details of which activities depend on which
- how long each activity will take (its **duration**).

You can assume for now that there are enough workers to complete each task as it is scheduled. You will meet problems with limited labour resources later.

This information is recorded as a **precedence table** or **dependence table**.

e.g. Suppose you are planning to paint your bedroom.
It might involve these activities:

A Remove furniture
B Remove curtains
C Remove carpets
D Wash ceiling
E Wash walls
F Paint ceiling
G Paint walls
H Replace carpets
I Replace curtains
J Replace furniture

Some activities clearly depend on others – you couldn't remove the carpets until you had removed the furniture, for example.

Here is a possible precedence table, with likely durations.

Activity	Duration (minutes)	Depends on
A	15	–
B	10	–
C	15	*A*
D	20	*A*, *B*, *C*
E	30	*A*, *B*, *C*
F	80	*D*
G	140	*E*
H	20	*F*, *G*
I	15	*F*, *G*
J	15	*H*

This assumes you have friends to help you, so that for example the ceiling can be painted at the same time as the walls.

The next stage is to draw an activity network.
The main features of this are:

- each activity is represented by an arc (edge)
- the weight of the arc is the duration of the activity
- each node (vertex) represents the event of completing one or more activities. The nodes are numbered
- there is one start (source) node (number 0) and one finish (sink) node.

e.g. In the bedroom painting example, *A* and *B* can both start straight away. *A* must finish before *C* can start. *A*, *B* and *C* must be complete before anything else can happen.

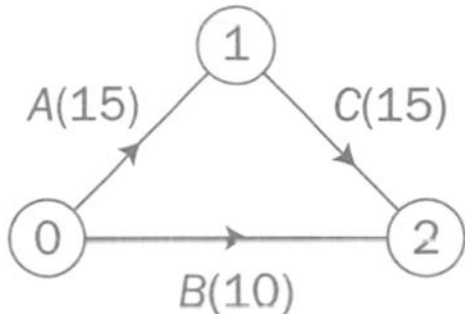

Node 2 is the **event** of *B* and *C* both being completed.

D and *E* can now start. *F* and *G* must follow *D* and *E* respectively. *F* and *G* must finish before anything else can happen.

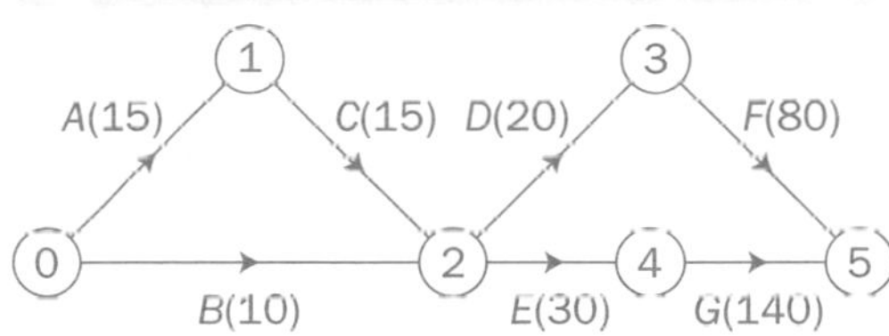

H and *I* can now start. *J* must follow *H*.

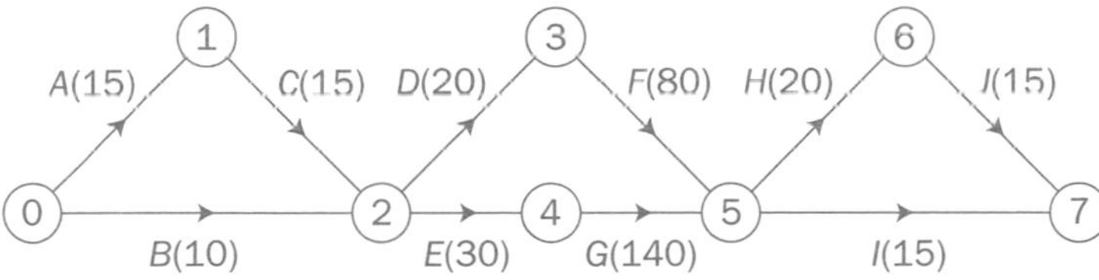

This is the complete activity network.

You can name an activity by its starting node (tail event) and finishing node (head event). E.g. Activity *E* above is (2, 4).

Identifying activities in this way is especially important if the data is to be computerised.

If this way of identifying activities is to work, two activities cannot have the same tail and head events.
To avoid this problem you may need to use a dummy activity, shown on the network as a dotted line.
A dummy activity has zero duration.

You should number nodes so that the head event of every activity has a higher number than its tail event.

D1

e.g. An activity network must **not** involve the following layout, because (3, 4) could refer to either activity *B* or activity *C*.

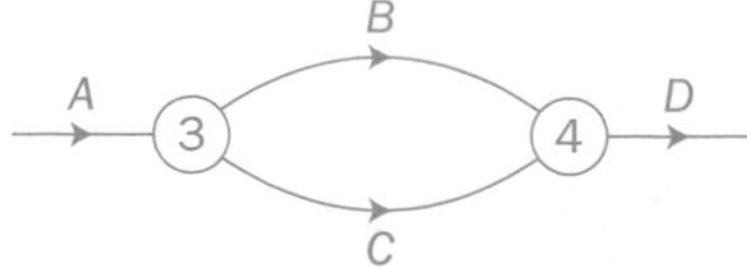

B and *C* here are **parallel events**.

You would redraw it as

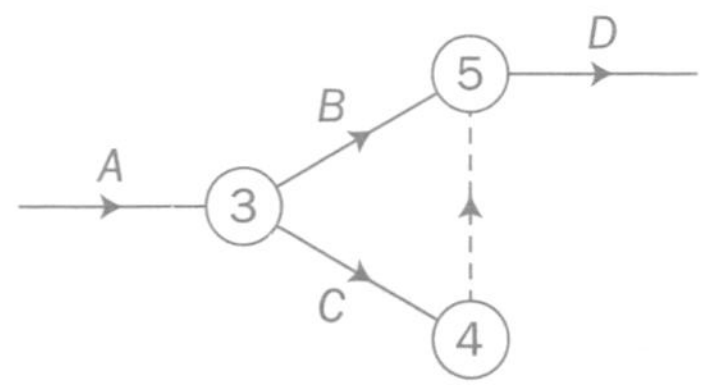

Activity (4, 5) is a **dummy activity**.

Activity *B* is now (3, 5) and activity *C* is (3, 4).

You need dummy activities in other cases. In particular, if two activities share only some of their predecessors, you need a dummy activity for the network to show the situation correctly.

e.g. Consider this precedence table.

Activity	Must be preceded by
A	–
B	A
C	A
D	B
E	B, C

The problem is to show that *E* depends on both *B* and *C*, while *D* depends only on *B*.

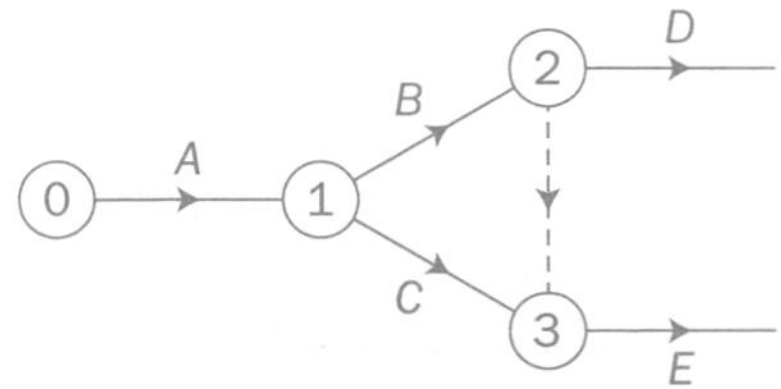

Here *E* depends on *C* and the dummy. As the dummy depends on *B*, *E* also depends on *B*.

EXAMPLE 1

Draw an activity network for the precedence table shown.

Activity	Duration (h)	Depends on
A	1	–
B	2	–
C	1	*A*
D	1	*A*
E	3	*B*
F	4	*C*
G	3	*C*
H	2	*C*, *D*, *E*
I	3	*F*, *G*

Notice that *F* and *G* depend only on *C* while *H* depends on *C*, *D* and *E*.
There must be an event '*C* has finished' and an event '*C*, *D* and *E* have finished'.
You need a dummy activity.

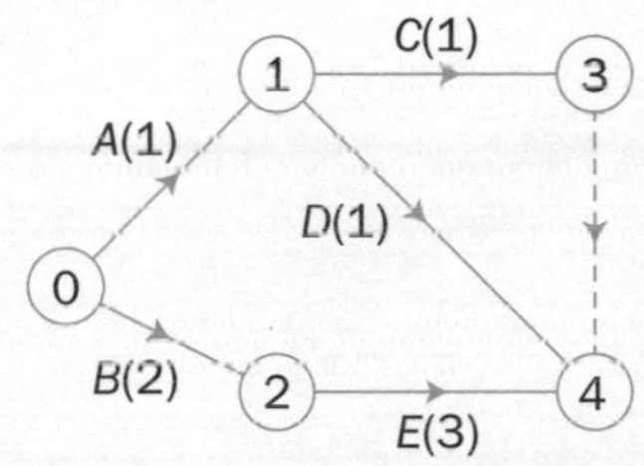

Node 3 is the event '*C* has finished'.
Node 4 is the event '*C*, *D* and *E* have finished'.

The other difficulty is that *F* and *G* are parallel events – they both depend on *C* and both precede *I*.
Another dummy is needed.

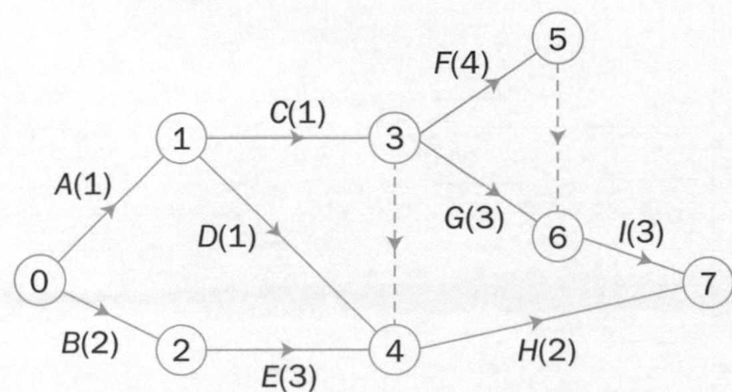

When drawing an activity network it is a good idea to make a rough sketch first. This helps you to design a clear layout – a graph in which no arcs cross if possible.

The answer is not unique – other layouts showing the same information are possible.

Have a go at finding some other layouts.

D1

Exercise 8.1

1 Draw an activity network for each of these precedence tables.

a

Activity	Duration	Must be preceded by
A	4	–
B	2	–
C	1	*A*
D	5	*A*
E	3	*B*
F	2	*C*, *E*

b

Activity	Duration	Must be preceded by
A	2	–
B	5	–
C	1	*A*
D	3	*B*
E	3	*B*
F	2	*C*
G	4	*D*, *E*

c

Activity	Duration	Must be preceded by
A	3	–
B	2	*A*
C	3	*A*
D	4	*A*
E	5	*B*, *C*
F	3	*C*, *D*
G	1	*D*

d

Activity	Duration	Must be preceded by
A	6	–
B	1	–
C	12	–
D	7	*A*, *B*
E	4	*D*
F	4	*D*
G	1	*E*, *F*
H	7	*F*
I	2	*C*, *G*
J	3	*I*

e

Activity	Duration	Must be preceded by
A	3	–
B	6	*A*
C	4	*A*
D	7	*A*
E	11	*B*
F	6	*B*
G	3	*C*, *D*, *F*
H	7	*D*
I	8	*E*
J	2	*E*
K	3	*G*, *I*
L	2	*G*, *H*, *I*
M	9	*K*, *L*

2 This is a list of tasks for making a wooden trolley for a child. The trolley will have wheels, a square wooden base, four wooden sides with a cord attached to one of them for towing the trolley.

Activity	Description	Depends on
A	Purchase wood	
B	Purchase wheels	
C	Purchase cord	
D	Cut out sides	
E	Cut out base	
F	Attach sides to base	
G	Attach wheels to base	
H	Attach cord	

Copy and complete the table. Hence draw an activity network for the project.

3 Draw up the precedence table corresponding to the activity network shown.
Label the activities with their tail and head events, e.g. $A(0, 1)$.

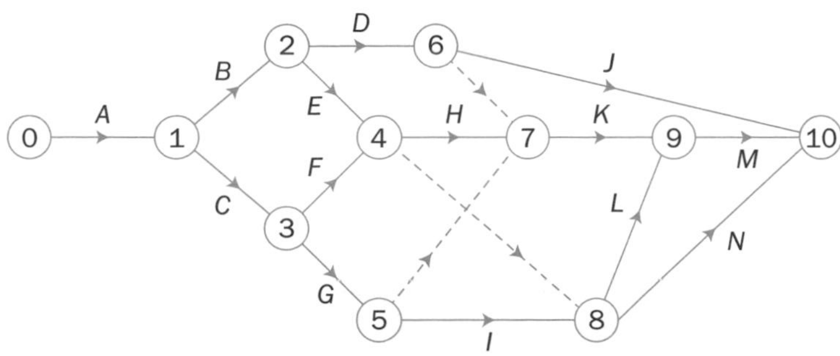

8.2 Finding the critical path

Each event (node) in a network represents the completion of the activities leading into it and the start of the activities leading out of it.

For each event i you need to know

- what is the earliest time by which you could complete all the incoming tasks? This is the **earliest event time** (e_i)
- what is the latest time at which you could start the outgoing tasks without increasing the overall length of the project? This is the **latest event time** (l_i).

You record these times at each node as | e_i | l_i |

To find the earliest event times you make a **forward pass** through the network. You start at the source node and work forward, recording the e_i values.
The source node has an earliest event time $e_0 = 0$.

e.g.

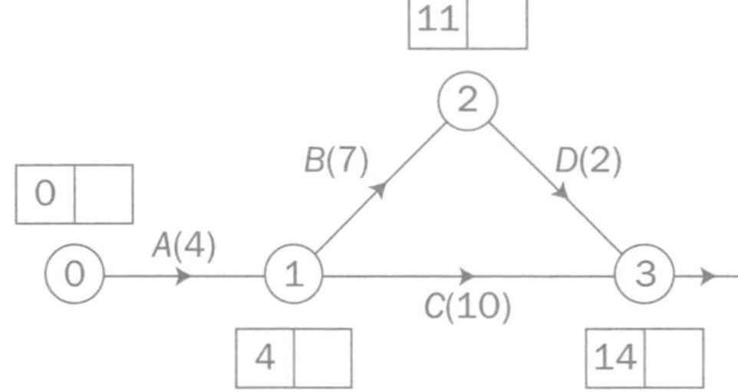

For this part-network you have

$e_0 = 0$

Event 1 has just one incoming activity, *A*.

$e_1 =$ (earliest start time for *A*) + (duration of *A*)

$= e_0 + 4 = 4$

Event 2 has just one incoming activity, *B*.

$e_2 =$ (earliest start time for *B*) + (duration of *B*)

$= e_1 + 7 = 11$

Event 3 has two incoming activities, *C* and *D*.

The earliest *C* can finish is e_1 + (duration of *C*)

$= 4 + 10 = 14$

The earliest *D* can finish is e_2 + (duration of *D*)

$= 11 + 2 = 13$

C and *D* must both be completed, so $e_3 = 14$.

If an event has several incoming activities, its earliest event time is the maximum of the times by which those activities can be completed.

For an event j with several incoming activities (i, j) you have

$$e_j = \text{maximum}\{e_i + \text{duration of } (i, j)\}$$

Completing the forward pass tells you how long the whole project will take.

The earliest event time for the finish event (sink node) is the minimum project duration.

To find the latest event times you then make a backward pass through the network. You start at the sink node and work back, recording the l_i values. The sink node has a latest event time equal to the project duration.

e.g. Suppose you have completed a forward pass and the final few nodes are as shown.

The project duration is 65.

You now make a backward pass.

l_{20} = project duration – 65

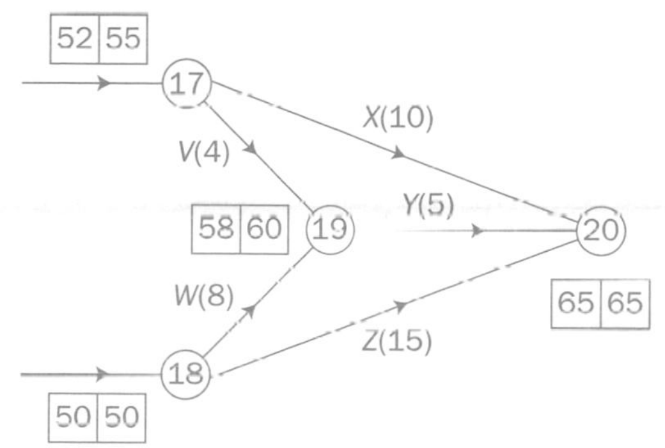

Event 19 has just one outgoing activity, Y.

l_{19} = (latest finish time for Y) – (duration of Y) = $l_{20} - 5 = 60$

Event 18 has two outgoing activities, W and Z.

The latest W can start is l_{19} – (duration of W) = $60 - 8 = 52$
The latest Z can start is l_{20} – (duration of Z) = $65 - 15 = 50$
To allow both of these, the latest event time is $l_{18} = 50$

Event 17 has two outgoing activities, V and X.

The latest V can start is l_{19} – (duration of V) = $60 - 4 = 56$
The latest X can start is l_{20} – (duration of X) = $65 - 10 = 55$
To allow both of these, the latest event time is $l_{17} = 55$

If an event has several outgoing activities, its latest event time is the minimum of the times by which those activities must start.

For an event i with several outgoing activities (i, j) you have

$$l_i = \text{minimum}\{l_j - \text{duration of } (i, j)\}$$

For some events $e_i < l_i$. A delay between completing the incoming activities and starting one or more outgoing activities will not affect the project duration. The event has slack.

The slack, s_i, for an event i is given by

$$s_i = l_i - e_i$$

If $e_i = l_i$, then event i has no slack and is a critical event. Any delay at critical events will increase the overall project duration.

An event i is critical if $e_i = l_i$

The time available for an activity (i, j) is the gap between its earliest possible start time and its latest possible finishing time.

Time available for activity $(i, j) = l_j - e_i$

In some cases the time available will be greater than the duration of the activity. The activity has float.

The total float, $F(i, j)$, for activity (i, j) is given by

$$F(i, j) = (l_j - e_i) - \text{duration of } (i, j)$$

An activity with zero float is a critical activity. An increase in its duration, or delay in starting it, will increase the overall project duration.

An activity joining critical events is not necessarily a critical activity.

An activity (i, j) is critical if $F(i, j) = 0$

The critical activities in a network form the longest path from the source to the sink. This is the critical path.

Critical path algorithm

Step 1 Perform a forward pass, recording earliest event times e_i

Step 2 Perform a backward pass, recording latest event times, l_i

Step 3 Calculate the total float $F(i, j)$ for each activity (i, j)

Step 4 Activities for which $F(i, j) = 0$ form the critical path

D1

EXAMPLE 1

Perform a forward and backward pass on the network shown. Hence find the project duration, the float for each activity and the critical path. Durations are in days.

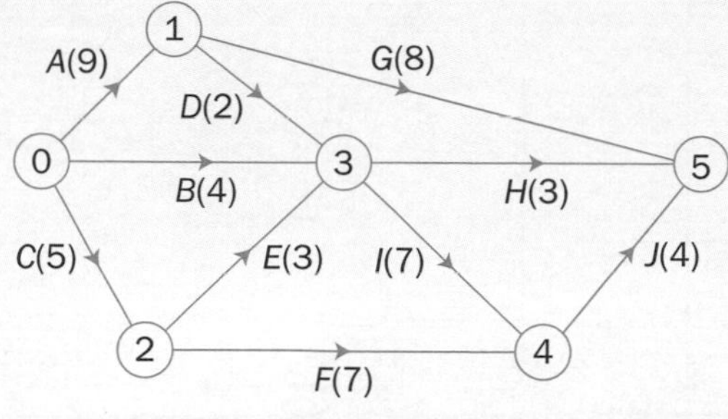

First perform a forward pass:

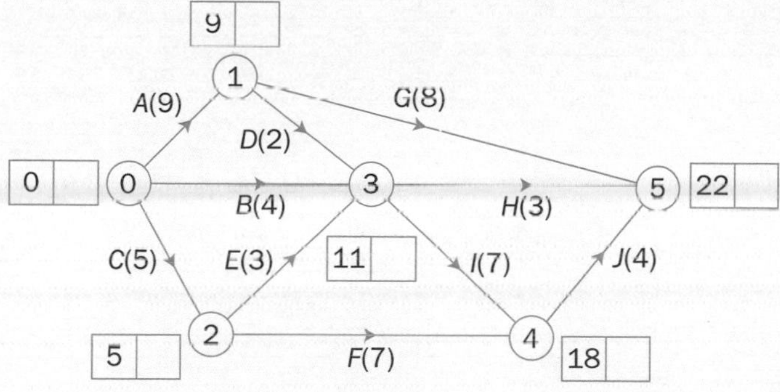

You should check that you can see how these event times were calculated.

The project duration is 22 days.

Next perform a backward pass:

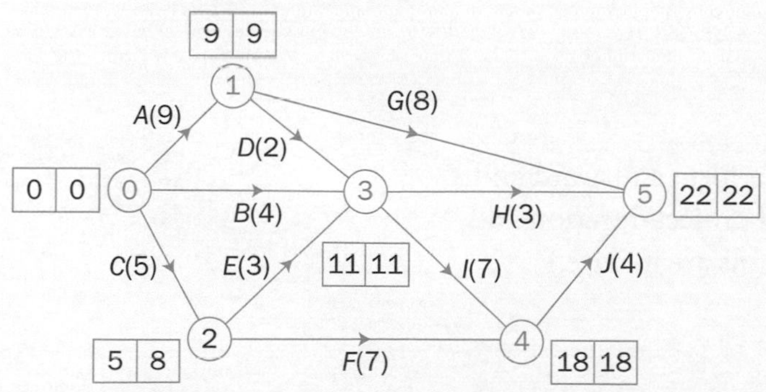

H joins two critical events (3 and 5), but *H* is not a critical activity.

The critical events are 0, 1, 3, 4 and 5, shown in blue.

Now find the total float for each activity:

e.g. For activity *H*, which is (3, 5), you have

$$F(3, 5) = l_5 - e_3 - \text{duration of } (3, 5)$$
$$= 22 - 11 - 3 = 8$$

So activity *H* has a total float of 8.

Example 1 is continued on the next page.

D1

EXAMPLE 1 (CONT.)

Show the floats in a table:

Activity (i, j)	Duration	e_i	l_j	Total float $F(i, j)$
A(0, 1)	9	0	9	**0**
B(0, 3)	4	0	11	7
C(0, 2)	5	0	8	3
D(1, 3)	2	9	11	**0**
E(2, 3)	3	5	11	3
F(2, 4)	7	5	18	6
G(1, 5)	8	9	22	5
H(3, 5)	3	11	22	8
I(3, 4)	7	11	18	**0**
J(4, 5)	4	18	22	**0**

The critical activities are A, D, I and J.
The critical path is shown below.

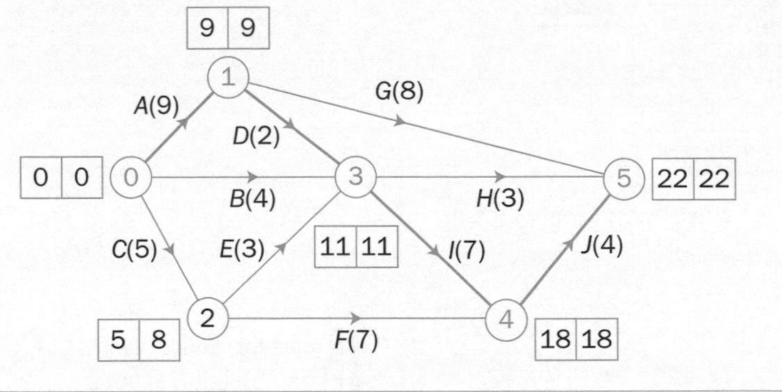

There may be more than one critical path. If activity G had duration 13 then it, too, would be a critical activity.

D1

Exercise 8.2

1 For each of these activity networks, perform a forward and a backward pass. Hence find the overall project duration and identify the critical activities. (Durations are in days.)

a

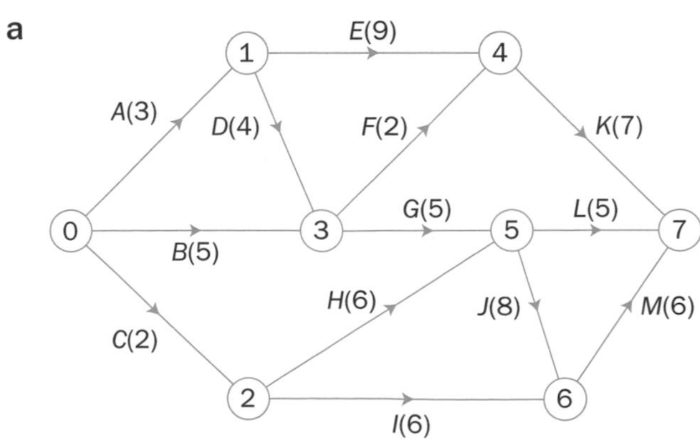

b

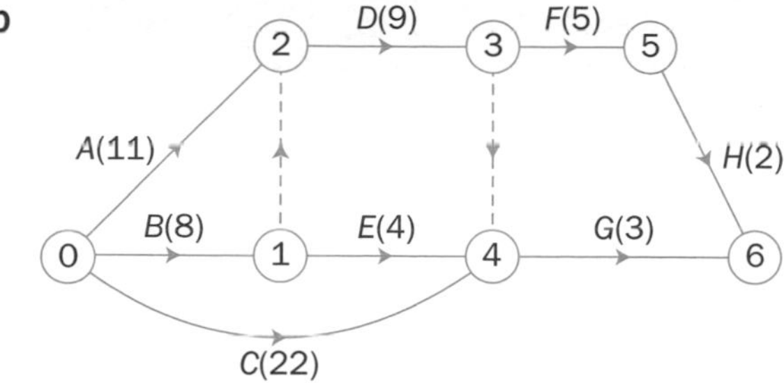

c

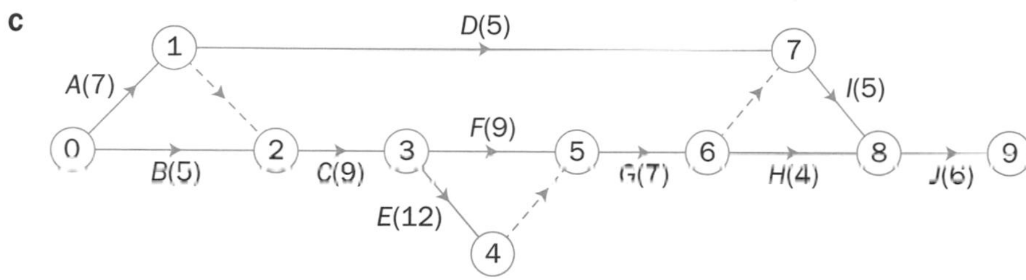

d

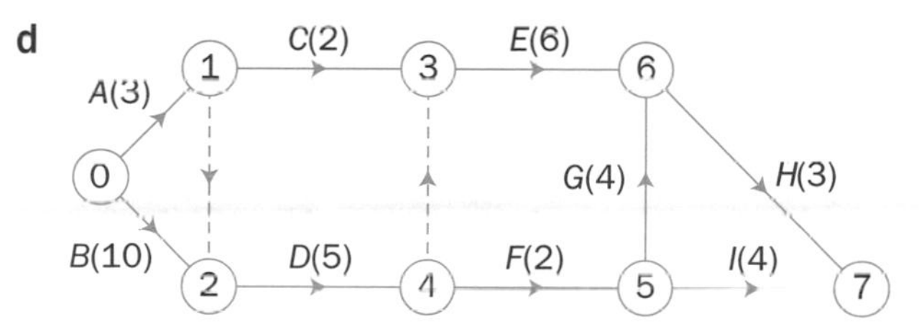

2 Determine the critical activities for the networks you drew in Exercise 8.1 question 1.

3 For the project shown in the table

a find the project duration

b determine the critical activities

c find the total float for those activities which are not critical.

Activity	Duration (hours)	Preceded by
A	5	–
B	4	–
C	6	–
D	8	A
E	4	A, B, C
F	7	C
G	3	D
H	6	D, E
I	8	F, G, H

8.3 Gantt charts

With non-critical activities, you have some flexibility about when to start them.

For an activity (i, j), you have already seen how to find its earliest possible start time (e_i) and its latest possible finish time (l_j).

You can also find its latest possible start time and its earliest possible finish time.

> For an activity (i, j)
>
> Earliest finish time $= e_i +$ duration of (i, j)
>
> Latest start time $= l_j -$ duration of (i, j)

e.g.

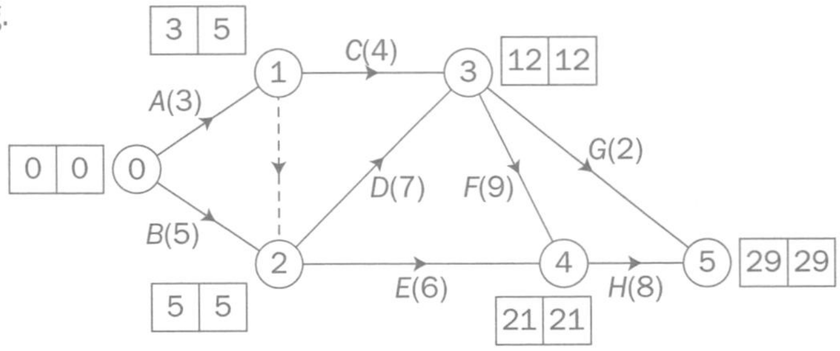

The table shows the complete information for this network.

For example, for activity C:

Earliest start time $= e_1 = 3$

Earliest finish time $= 3 +$ duration of $C = 7$

Latest finish time $= l_3 = 12$

Latest start time $= 12 -$ duration of $C = 8$

Activity	Duration (days)	Start time		Finish time		Float
		Earliest	Latest	Earliest	Latest	
A	3	0	2	3	5	2
B	5	0	0	5	5	0
C	4	3	8	7	12	5
D	7	5	5	12	12	0
E	6	5	15	11	21	10
F	9	12	12	21	21	0
G	2	12	27	14	29	15
H	8	21	21	29	29	0

Check that you can see how these values were obtained.

The float can be calculated as the difference between the early and late start times, or the early and late finish times.

You can now illustrate this information on a **Gantt chart** or **cascade chart**.

Activities are shown as bars against a time scale.
The critical activities are fixed, so you insert those first.

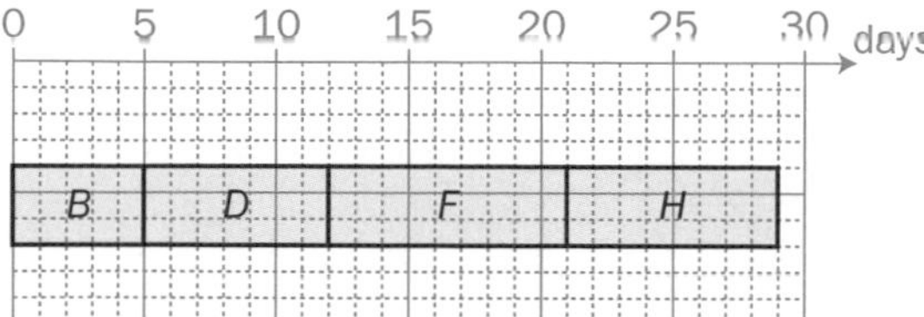

You place each non-critical activity at its earliest start time, and show the boundaries within which it can be moved.

- Activity *A* could start at 0 and has a latest finish of 5.
- Activity *C* has an earliest start of 3 and a latest finish of 12.
- Activity *E* has an earliest start of 5 and a latest finish of 21.
- Activity *G* has an earliest start of 12 and a latest finish of 29.

The finished Gantt chart looks like this:

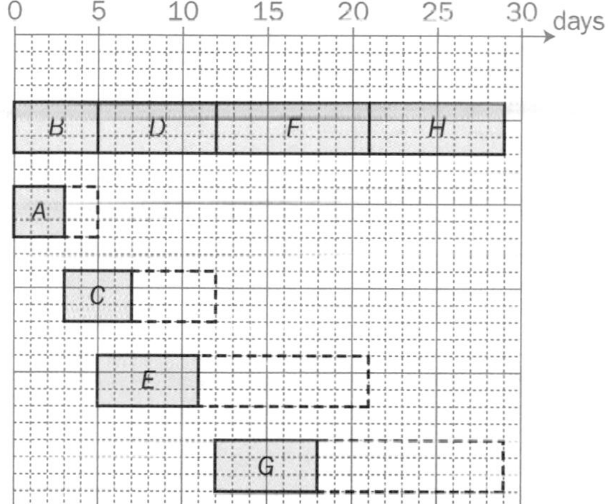

The Gantt chart shown was slightly simplified. Although *C* has 5 days float, 2 of these depend on when *A* starts. *C* has 2 days of dependent float and 3 days of independent float. In some books you will see this shown on the chart as a 'fence', indicating *C* cannot start until *A* is complete, as shown here. However, the present syllabus does not require you to do this.

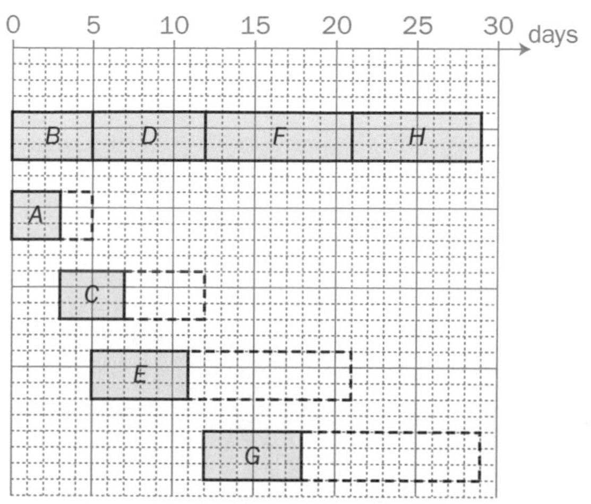

All of the float for *E* and *G* is independent float.

D1

Exercise 8.3

1 Copy and complete this table.
Hence draw a Gantt chart to show the project.

Activity	Duration (hours)	Start		Finish		Float
		Earliest	Latest	Earliest	Latest	
A	6	0			6	
B	8	0			15	
C	14	6			22	
D	12	6			18	
E	3	8			18	
F	4	18			22	
G	3	22			25	

2 Draw a Gantt chart for the activity network shown.
(This is the network from Exercise 8.2 question 1b – durations are in days.)

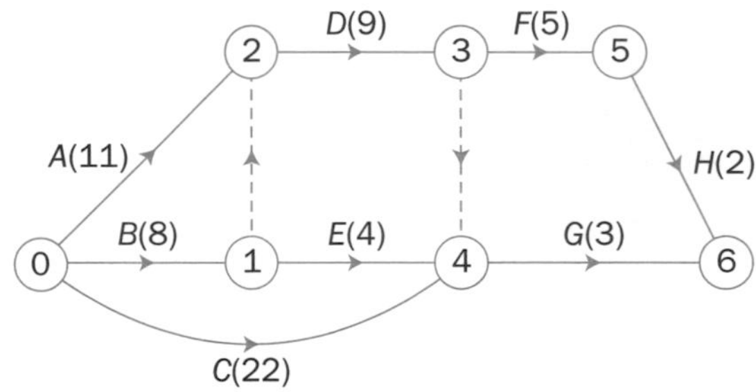

3 Draw a Gantt chart for the activity network shown.
(This is the network from Exercise 8.2 question 1c – durations are in days.)

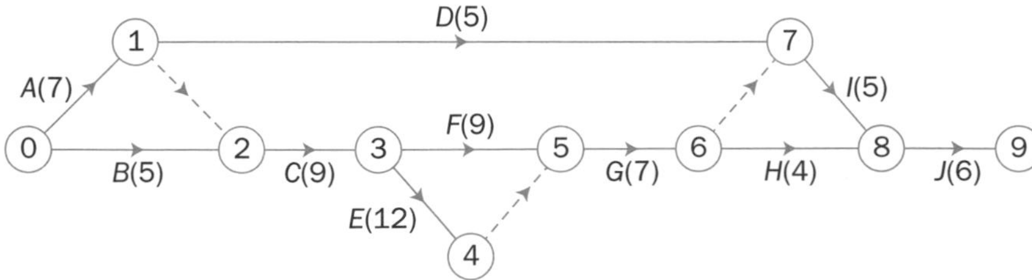

D1

4 Draw a Gantt chart for the activity network shown. (This is the network from Exercise 8.2 question 1d – durations are in days.)

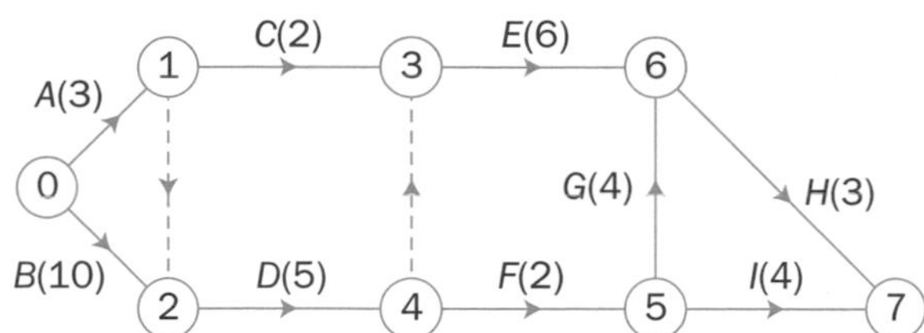

5

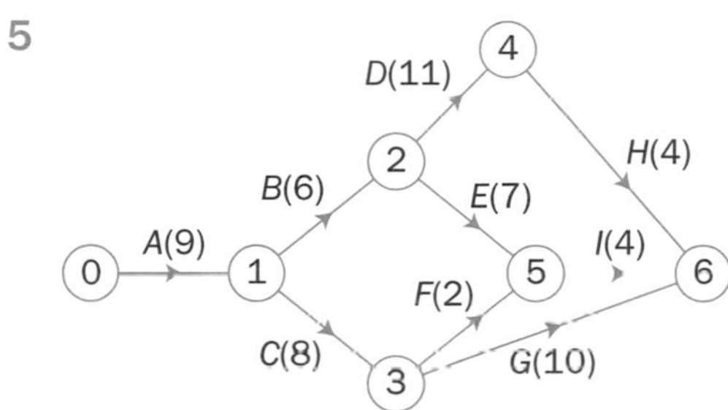

a Perform a forward and a backward pass on the activity network shown.
The durations are given in hours.

b Produce a table of earliest and latest start and finish times.

c Draw a cascade chart to show the project.

d It is possible to reduce the duration of activity B or activity E by up to 4 hours.
It costs £50 per hour to do so.
On which activity should the money be spent to reduce the overall project length?
What is the most that should be spent?

D1

8.4 Scheduling

The aim of scheduling is to allocate workers to activities so that the project is completed as quickly as possible. You may need to

- decide how many workers are required to complete the project in its minimum time

or

- decide how long the project will take with the available number of workers.

To simplify things, you assume

- each activity needs one worker
- no worker can stand idle if there is an available activity
- once a worker starts an activity it will be continued until it is complete.

There is no perfect algorithm for scheduling. The following procedure is commonly used. It uses the idea of an imaginary project clock recording how long the project has been going.

Scheduling procedure

The project clock starts at 0.

Step 1 If there is a free worker and an activity which can be started, assign the worker to the activity.
If there is a choice of activities, choose the most critical one.
Repeat this until no worker is free or no activity is available.

Step 2 If all activities have been assigned, stop.
Otherwise, advance the project clock until there is a free worker and an activity which can be started, and go to Step 1.

The **most critical activity** is the one with the smallest latest starting time.

EXAMPLE 1

The diagram shows the activity network of a project. Durations are in days.

a How many workers are needed if every activity is to start at its earliest time?

b Show that the project can be scheduled for three workers without increasing the overall project duration.

c How long would the project take if only two workers were available?

Activity	Duration (days)	Start		Finish		Float
		Earliest	Latest	Earliest	Latest	
A	4	0	0	4	4	0
B	7	4	4	11	11	0
C	12	4	5	16	17	1
D	6	11	17	17	23	6
E	4	11	20	15	24	9
F	8	11	11	19	19	0
G	2	16	17	18	19	1
H	6	16	21	22	27	5
I	5	19	19	24	24	0
J	4	17	23	21	27	6
K	3	24	24	27	27	0

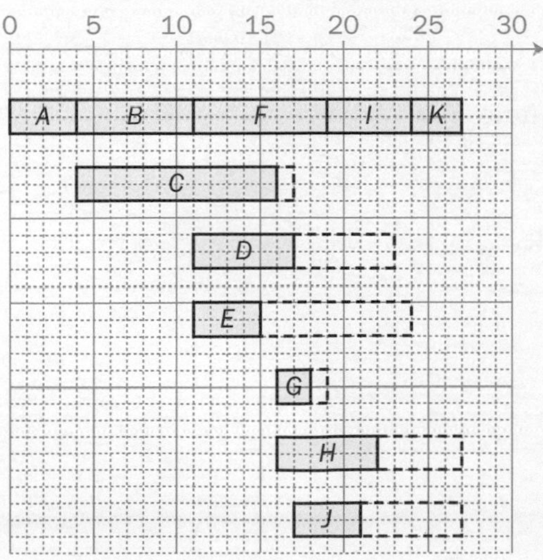

The time analysis of the project is shown in the table and illustrated by the Gantt chart.

Example 1 is continued on the next page.

D1

EXAMPLE 1 (CONT.)

a Assume there is an unlimited supply of workers. (W1 stands for Worker 1.)

Step 1 Clock set to zero. Give W1 activity *A*. No more assignments possible.

Step 2 Advance clock to 4.

Step 1 W1 becomes free. *B* and *C* can start. Give *B* to W1, *C* to W2. No more possible.

At this stage you have

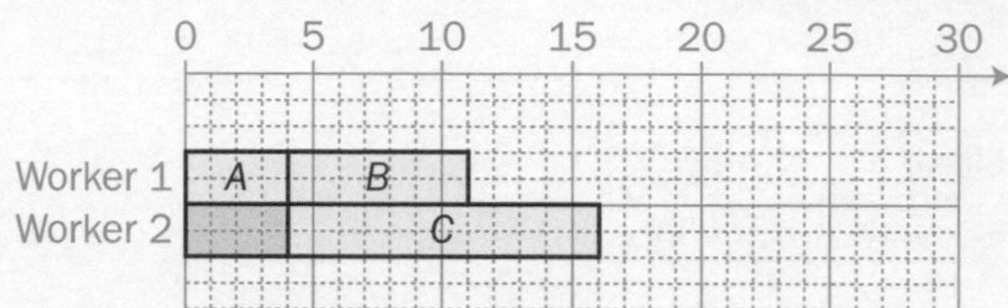

Step 2 Advance clock to 11.

Step 1 W1 becomes free. *F*, *D* and *E* can start. Give *F* to W1, *D* to W3, *E* to W4. No more possible.

You now have

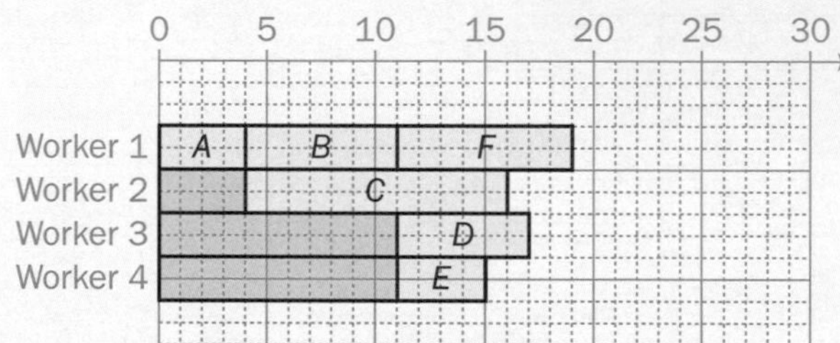

Step 2 Advance clock to 15. W4 becomes free but no activity can start. Advance clock to 16.

Step 1 W2 becomes free. *G* and *H* can start. Give *G* to W2, *H* to W4. No more possible.

Step 2 Advance clock to 17.

Step 1 W3 becomes free. *J* can start. Give *J* to W3. No more possible.

You now have

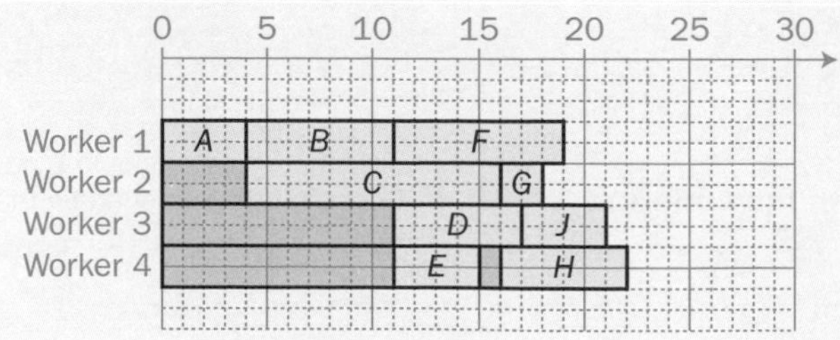

Example 1 is continued on the next page.

D1

EXAMPLE 1 (CONT.)

Step 2 Advance clock to 19.
Step 1 W1 becomes free. *I* can start.
Give *I* to W1. No more possible.
Step 2 Advance clock to 24.
Step 1 All workers now free. *K* can start.
Give *K* to W1.
Step 2 All activities now assigned, so stop.

Your final schedule is

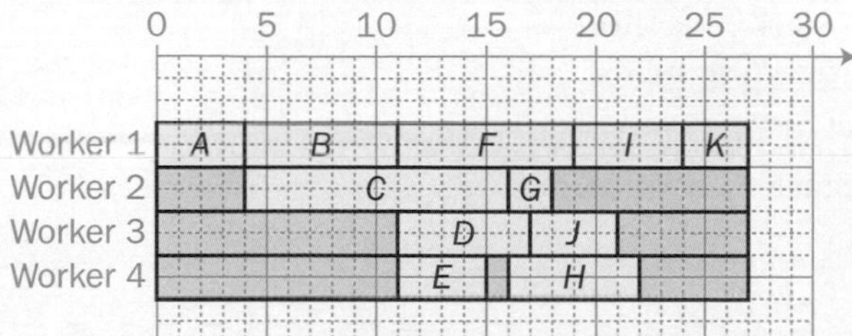

So four workers are needed if every activity starts at its earliest time.

b With three workers available, some activities will require a later starting time. The schedule is the same as part a until the clock reaches 11:

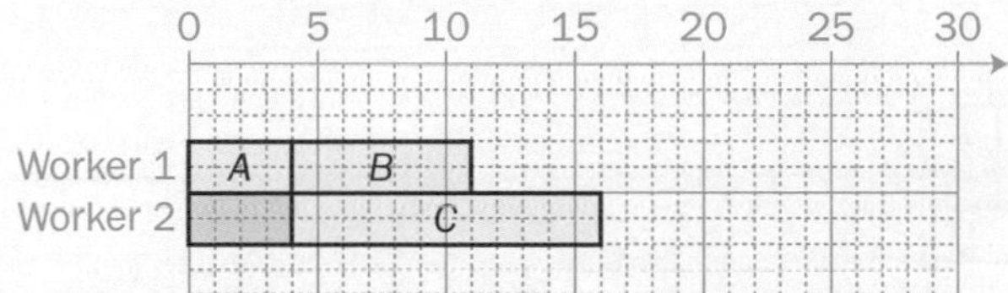

Step 2 Advance clock to 11.
Step 1 W1 becomes free. *F*, *D* and *E* can start.
F is a critical activity, so give *F* to W1.
D (latest start time 17) is more critical than *E* (latest start time 20), so give *D* to W3.
No more possible.
Step 2 Advance clock to 16.
Step 1 W2 becomes free. *E*, *G* and *H* can start.
G (latest start time 17) is the most critical, so give *G* to W2. No more possible.
Step 2 Advance clock to 17
Step 1 W3 becomes free. *E*, *H* and *J* can start.
E is the most critical, so give *E* to W3.
No more possible.

Example 1 is continued on the next page.

D1

EXAMPLE 1 (CONT.)

At this stage you have

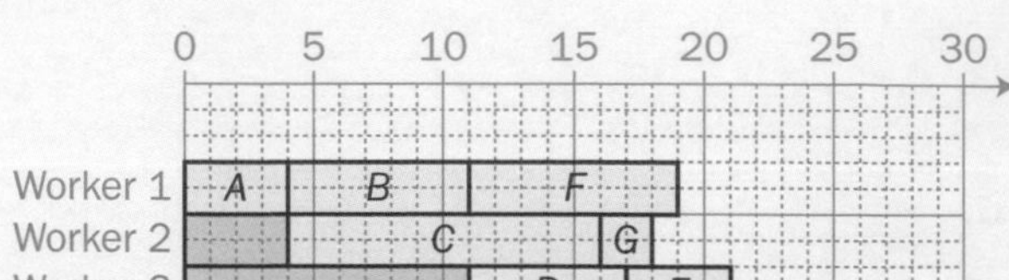

Step 2 Advance clock to 18.

Step 1 W2 becomes free. *H* and *J* can start. *H* is more critical, so give *H* to W2. No more possible.

Step 2 Advance clock to 19.

Step 1 W1 becomes free. *I* and *J* can start. *I* is critical, so give I to W1. No more possible.

Step 2 Advance clock to 21.

Step 1 W3 becomes free. *J* can start. Give *J* to W3. No more possible.

Step 2 Advance clock to 24.

Step 1 W1 becomes free. *K* can start. Give *K* to W1.

Step 2 All activities now assigned, so stop.

The complete schedule is

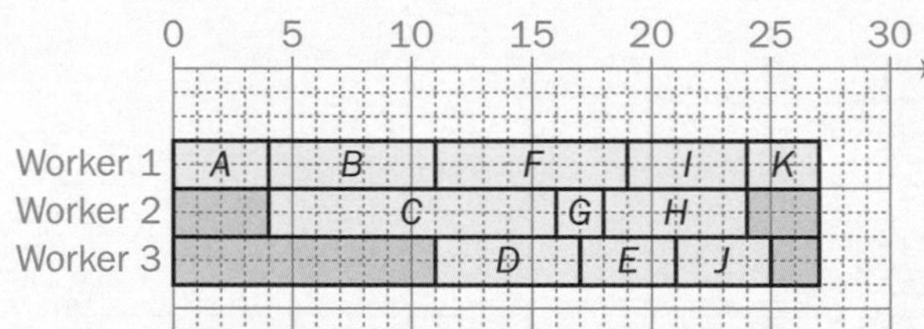

This schedule corresponds to adjustments in the Gantt chart as shown.

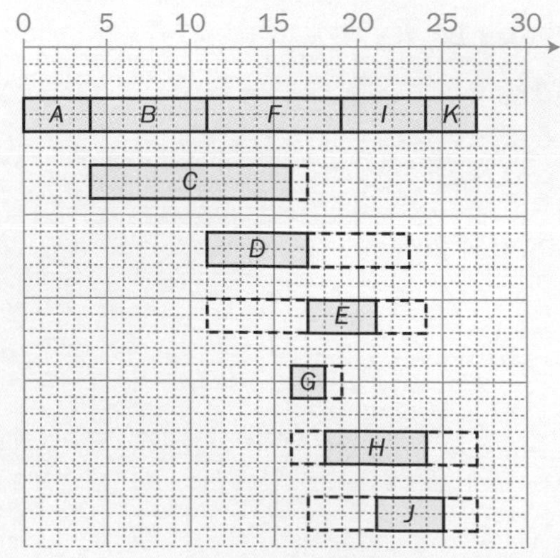

Part **c** of Example 1 is answered on the next page.

D1

EXAMPLE 1 (CONT.)

c If only two workers are available, the project must take longer. If the first worker does all the critical activities, the second does the non-critical activities *C*, *D*, *E*, *G*, *H* and *J*, which take a total of 35 days. This is longer than the current project duration of 27 days.

The scheduling procedure is as follows.

Step 1 Clock set to zero. Give *A* to W1.
No more possible.
Step 2 Advance clock to 4.
Step 1 W1 becomes free. *B* and *C* can start.
Give *B* to W1, *C* to W2. No more possible.

At this stage you have

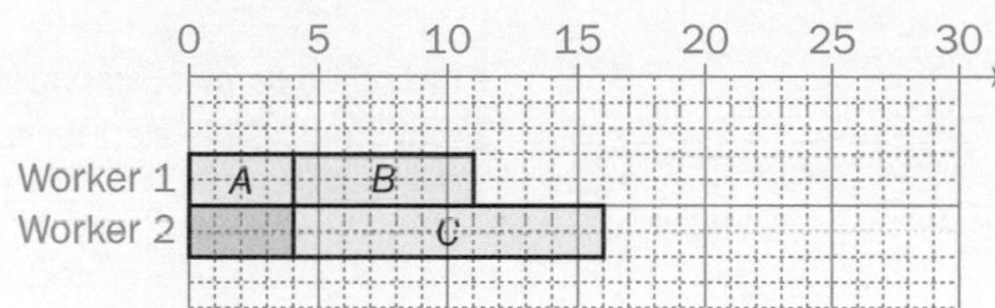

Step 2 Advance clock to 11.
Step 1 W1 becomes free. *F*, *D* and *E* can start.
Give *F* (critical) to W1. No more possible.
Step 2 Advance clock to 16.
Step 1 W2 becomes free. *D*, *G* and *H* can start. *D* and *G* are equally critical. Give *D* to W2.
No more possible.
Step 2 Advance clock to 19.
Step 1 W1 become free. *E*, *G*, *H*, *I* and *J* can start. *G* is most critical, so give *G* to W1. No more possible.

At this stage all the activities are available to start. The schedule looks like this.

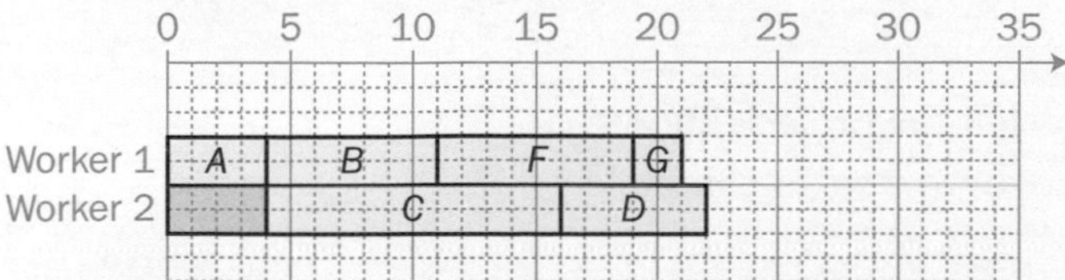

Continuing to assign workers to activities as the workers become free, the final schedule is

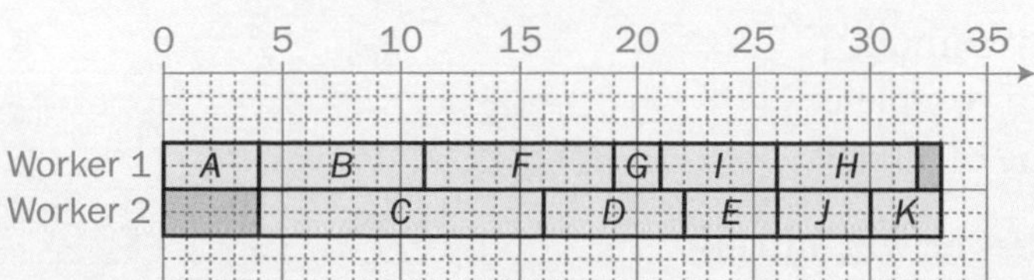

The project will now take a total of 33 days.

DI

In Example 1 there is only one day difference in the finishing times of the two workers, so rearranging the later activities will not improve the project time. In some cases you may gain some advantage in this way.

e.g.

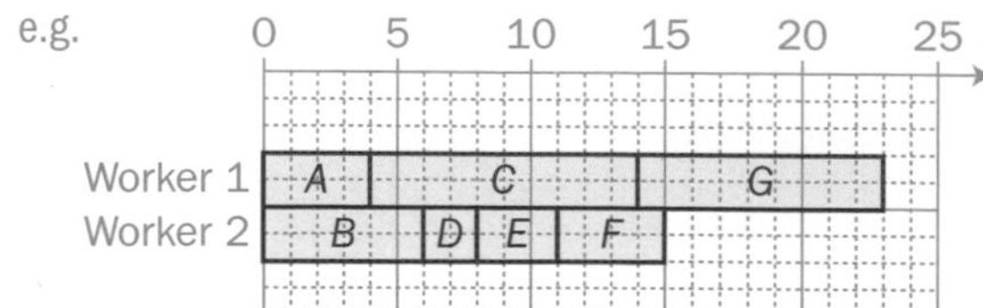

This schedule gives a project time of 23 days. This could be improved to 19 days as follows.

The scheduling procedure does not always give this optimal solution.

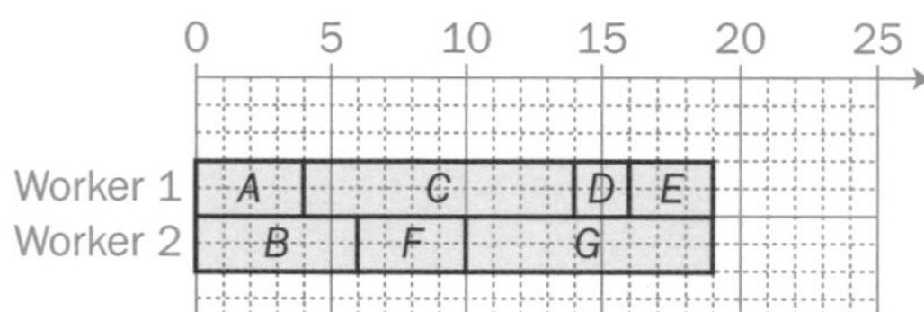

The schedule must still be in line with the possible start times for the activities.

Exercise 8.4

1 In Exercise 8.3 question 1 you completed a table analysing a project with a total project time of 25 hours.

- **a** Show that three workers are needed if every activity is to start within its allowed start times. Draw up a schedule for these workers.
- **b** In fact only two workers are available. Activity *G* cannot start until activity *C* is complete. Draw up a schedule for these workers and find the revised project time.

2 **a** A team of three workers is assigned to the project shown in Exercise 8.3 question 2. Draw up a schedule for the workers to complete the project on time. For how long will there be workers standing idle?

- **b** After 12 days one of the workers left the project. Show that it is possible for the remaining two workers to complete the project on time.

3 The diagram shows the activity network from Exercise 8.3 question 5. The durations are in days.

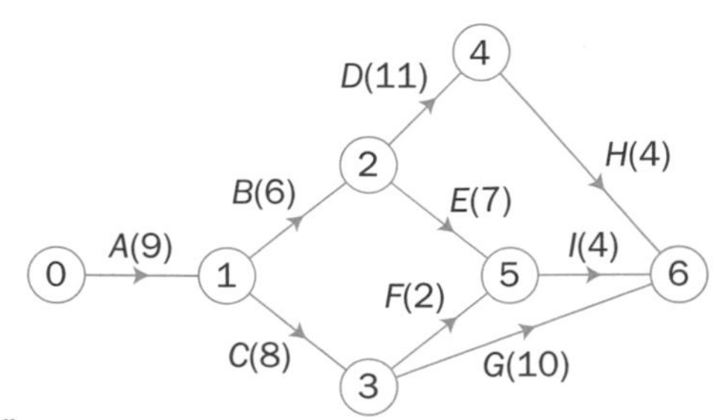

- **a** Draw up a schedule using the minimum number of workers necessary to complete the project in its minimum time of 30 days.
- **b** The project manager decides that he will initially assign two workers to the project and more at a later stage in order to complete the project on time. At what point should the extra worker(s) be introduced?

4 The diagram shows the activity network for a project. Durations are in weeks.

a Analyse the project to find its minimum completion time. Find the earliest and latest start and finish times for each activity.

b Find the sum of the durations of all the activities. Explain how this shows that it is not possible for two workers to complete the project in its minimum time.

c Show that by starting either A or B late it is possible to schedule the project for three workers.

d If just two workers are available, find the minimum time its will take to complete the project.

5 The diagram shows the activity network from Exercise 8.2 question **1a**. Durations are in days. Schedule this project for a team of three workers.

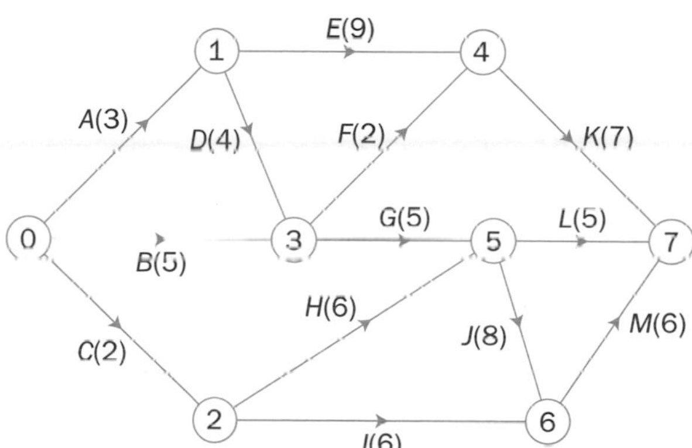

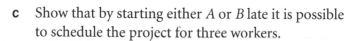

6

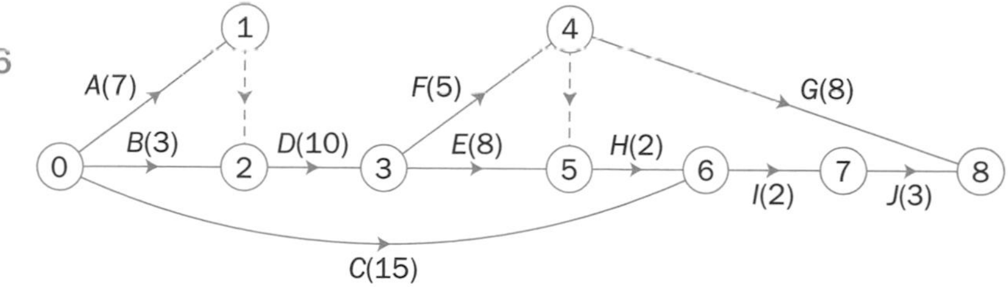

The diagram shows the activity network for a project. Durations are in weeks.

a Find the minimum completion time for this project.

b Show that it is possible for two workers to complete the project in this time.

c The project manager finds that activity E could be shortened to 7 weeks. Explain the implications of such a move for
i the minimum project time ii the schedule.

Review 8

1 The precedence table for activities involved in manufacturing a toy is shown.

Activity	Preceding activity
A	–
B	–
C	–
D	A
E	A
F	B
G	B
H	C, D, E, F
I	E
J	E
K	I
L	I
M	G, H, K

a Draw an activity network, using activity on arc, and exactly one dummy, to model the manufacturing process.

b Explain briefly why it is necessary to use a dummy in this case. [(c) Edexcel Limited 2002]

2 a Construct an activity network for the project in this table. Explain why you needed to use a dummy activity.

Activity	Must be preceded by	Duration (days)
A	–	10
B	–	8
C	A	5
D	B	5
E	C, D	4
F	C, D	4
G	E	6
H	E, F	2

b Find the earliest and latest times for the events in your network. Hence find

i the minimum completion time for the whole project

ii the critical activities.

c Draw a cascade chart for the project.

3 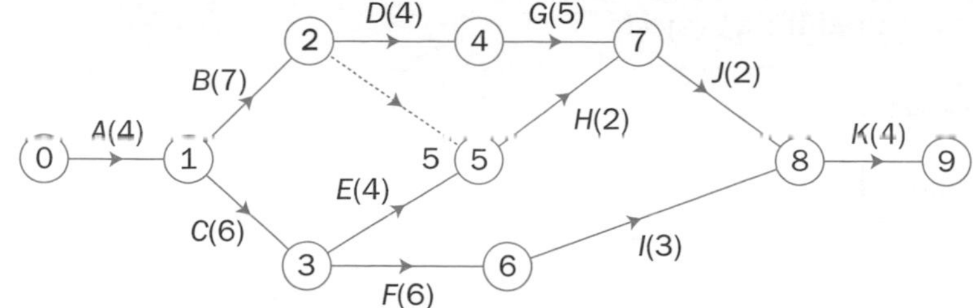

a The diagram shows the activity network for a project, with durations in days. Find the completion time for the project shown and state the critical activities.

b By how much could activity *E* overrun without affecting the overall completion time?

c The project manager is keen to complete the project more quickly, and finds that she can shorten activities *B*, *D* and *G* by a day each, at a cost of £100 per day. Advise her how much to spend.

4 a Draw an activity network described in this precedence table, using as few dummies as possible.

Activity	Must be preceded by
A	–
B	A
C	A
D	A
E	C
F	C
G	B, D, E, F
H	B, D, E, F
I	F, D
J	G, H, I
K	F, D
L	K

b A **different** project is represented by the activity network shown. The duration of each activity is shown in brackets.

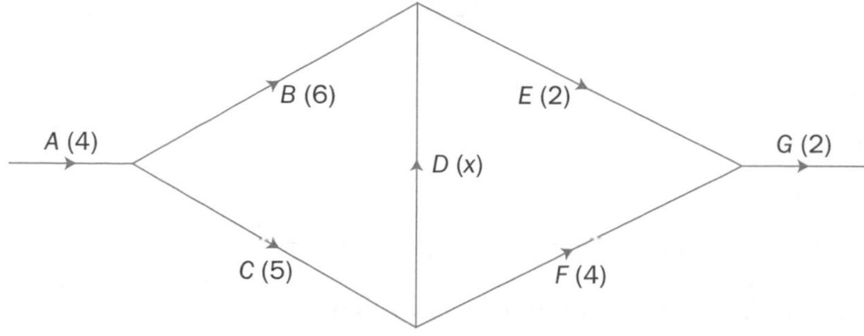

Find the range of values of *x* that will make *D* a critical activity. [(c) Edexcel Limited 2003]

D1

5 The table shows the activities required for a project.

Activity	Must be preceded by	Duration (days)
A	–	1
B	*A*	3
C	*A*	5
D	*B*, *C*	7
E	*D*	2
F	*C*	1
G	*D*, *F*	3
H	*D*, *F*	5
I	*E*, *G*, *H*	3

a Draw an activity network for this project.

b Calculate the total completion time for the project, and state the critical activities.

c Calculate the float for each of the non-critical activities.

d Draw a Gantt chart for the project.

e Assuming that each activity only requires one worker, draw up a schedule to show that the project can be completed in time by just two workers.

6

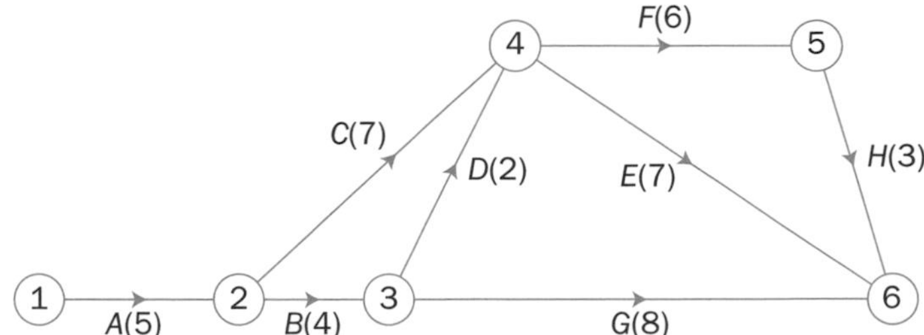

A project is modelled by the activity network shown. The activities are represented by the edges. The number in brackets on each edge gives the time, in days, taken to complete the activity.

a Calculate the early time and the late time for each event.

b Hence determine the critical activities and the length of the critical path.

c Obtain the total float for each of the non-critical activities.

d Draw a cascade (Gantt) chart showing the information obtained in parts **b** and **c**.

Each activity requires one worker. Only two workers are available.

e Draw up a schedule and find the minimum time in which the two workers can complete the project. [(c) Edexcel Limited 2002]

7

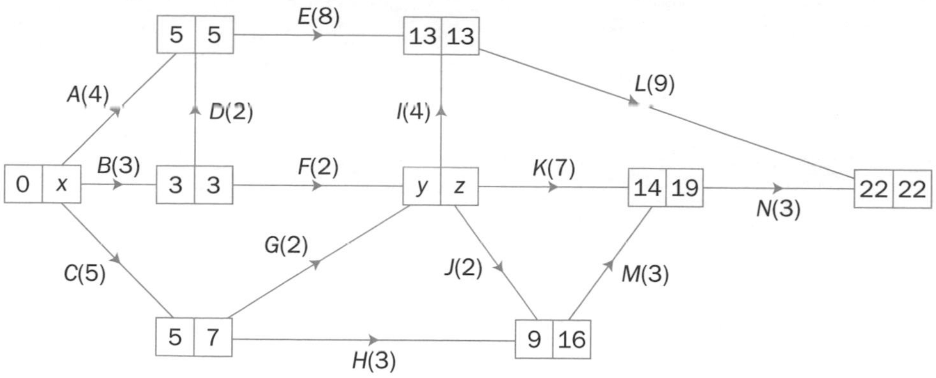

A trainee at a building company is using critical path analysis to help plan a project. The trainee's activity network is shown. Each activity is represented by an arc and the number in brackets on each arc is the duration of the activity, in hours.

a Find the values of x, y and z.

b State the total length of the project and list the critical activities.

c Calculate the total float time on **i** activity N **ii** activity H.

The trainee's activity network is checked by the supervisor who finds a number of errors and omissions in the diagram. The project should be represented by this precedence table.

Activity	Must be preceded by	Duration
A	–	4
B	–	3
C	–	5
D	B	2
E	A, D	8
F	B	2
G	C	2
H	C	3
I	F, G	4
J	F, G	2
K	F, G	7
L	E, I	9
M	H, J	3
N	E, I, K, M	3
P	E, I	6
Q	H, J	5
R	Q	7

d By adding activities and dummies, amend the diagram so that it represents this precedence table. (The durations shown on the trainee's diagram are correct.)

e Find the total time needed to complete this project.

[(c) Edexcel Limited 2004]

8 Exit

Summary

Refer to

- A precedence table lists the activities involved in a project, their durations and details of which activities must precede which. 8.1
- You can model a project using a directed network. 8.1
- The project timetable is analysed as follows:
  - make a forward pass to find the earliest event time (e_i) for each event i. The earliest event time for the finish node is the project completion time
  - make a backward pass to find the latest event time (l_i)
  - event e_i is a critical event if $e_i = l_i$
  - for an activity (i, j), the float $F(i, j) = l_j - e_i -$ duration of (i, j)
  - activity (i, j) is a critical activity if $F(i, j) = 0$. 8.2
- You can illustrate the project timetable using a Gantt chart (cascade chart). 8.3
- Scheduling is the systematic allocation of workers to activities. You can illustrate the schedule for a project using a chart in which activities are represented by bars in relation to a time scale. 8.4

Links

Critical path analysis can be applied to all examples of project management in real life. These can range from refitting a cruise liner, to implementing a new software or to activities such as preparing for a holiday.

Revision 2

1 A college wishes to staff classes in French (F), German (G), Italian (I), Russian (R) and Spanish (S). Five language teachers are available – Mr Ahmed (A), Mrs Brown (B), Ms Corrie (C), Dr Donald (D) and Miss Evans (E). The languages they can teach are shown in the table.

Mr Ahmed (A)	French and German
Mrs Brown (B)	French and Italian
Ms Corrie (C)	Russian
Dr Donald (D)	Russian and Spanish
Miss Evans (E)	French and Spanish

At first three teachers were allowed to chose their preferences. Their choices were:

Mr Ahmed (A) – French (F)
Dr Donald (D) – Russian (R)
Miss Evans (E) – Spanish (S)

a Draw a bipartite graph to show the information in the table. Indicate the three preferences chosen in a distinctive way.

b Using your answer to part a as the initial matching apply the maximum matching algorithm to obtain a complete matching. Alternating paths and the final matching should be stated. [(c) Edexcel Limited 2002]

2 Six workers A, B, C, D, E and F are to be matched to six tasks 1, 2, 3, 4, 5 and 6. The table shows the tasks that each worker is able to do.

Worker	Tasks
A	2, 3, 5
B	1, 3, 4, 5
C	2
D	3, 6
E	2, 4, 5
F	1

a Draw a bipartite graph to show this information.

Initially, A, B, D and E are allocated tasks 2, 1, 3 and 5 respectively.

b Starting from the given initial matching, use the matching improvement algorithm to find a complete matching, showing your alternating paths clearly.

3

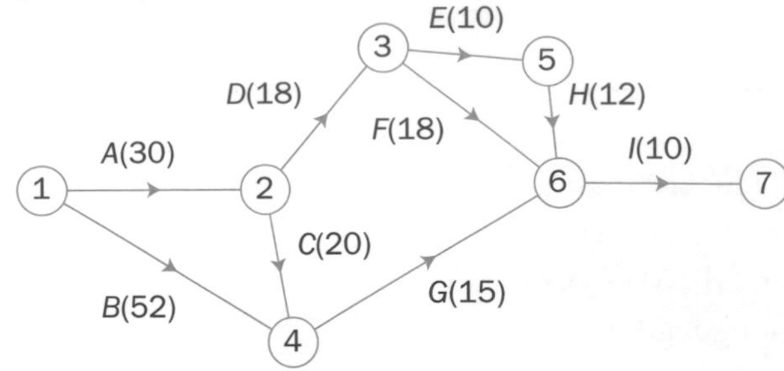

A building project is modelled by the activity network shown. The activities involved in the project are represented by the arcs. The numbers in brackets on each arc give the time, in days, taken to complete the activity.

a On a copy of the diagram, calculate the earliest and latest event times.

b Hence write down the critical activities and the length of the critical path.

c Obtain the total float for each non-critical activity.

d Draw a Gantt chart to show the information found in parts **b** and **c**.

Given that each activity requires one worker,

e draw up a schedule to determine the minimum number of workers needed to complete the project in the critical time.

Due to unforeseen circumstances, activity *C* takes 30 days rather than 20 days.

f Determine how this affects the length of the critical path and state the critical activities now.

[(c) Edexcel Limited 2002]

4 The precedence table for a certain project is as shown.

a Draw an activity network to show this project.

b Find the earliest and latest event times for your network.

c State the critical activities and the length of the critical path.

d Draw a cascade chart to illustrate the project.

e Each activity requires one worker. Find the minimum number of workers needed to complete the project on time.
Draw up a schedule for these workers.

Activity	Must be preceded by	Duration (days)
A	–	4
B	–	6
C	–	7
D	*A*	3
E	*A*	10
F	*B*, *D*	9
G	*C*	5
H	*C*	4
I	*E*, *F*, *G*	6
J	*F*, *G*	11
K	*H*	2
L	*E*, *F*, *G*	3

D1

5 A company produces two types of party bag, Infant and Junior. Both types of bag contain a balloon, a toy and a whistle. In addition the Infant bag contains 3 sweets and 3 stickers, and the Junior bag contains 10 sweets and 2 stickers.
The sweets and the stickers are produced in the company's factory. The factory can produce up to 3000 sweets per hour and 1200 stickers per hour. The company buys a large supply of balloons, toys and whistles.
Market research indicates that at least twice as many Infant bags as Junior bags should be produced.
Both types of party bag are sold at a profit of 15p per bag. All the bags are sold. The company wishes to maximise its profit.
Let x be the number of Infant bags produced and y be the number of Junior bags produced per hour.

a Formulate the above situation as a linear programming problem.

b Draw a graph to show the inequalities, indicating clearly the feasible region.

c Find the number of Infant bags and Junior bags that should be produced each hour and the maximum hourly profit. Make your method clear.

In order to increase the profit further, the company decides to buy additional equipment. It can buy equipment to increase the production of **either** sweets or stickers, but **not both**.

d Using your graph, explain which equipment should be bought, giving your reasoning.

The manager of the company does not understand why the balloons, toys and whistles have not been considered in the above calculations.

e Explain briefly why they do not need to be considered. [(c) Edexcel Limited 2006]

6 Two fertilisers are available, a liquid X and a powder Y. A bottle of X contains 5 units of chemical A, 2 units of chemical B and $\frac{1}{2}$ unit of chemical C, and costs £2. A packet of Y contains 1 unit of A, 2 units of B and 2 units of C, and costs £3.

A professional gardener makes her own fertiliser. She requires at least 10 units of A, at least 12 units of B and at least 6 units of C. She buys x bottles of X and y packets of Y. She wishes to minimise her total cost £T.

a Formulate this situation as a linear programming problem.

b Draw a graph showing the constraints and label the feasible region.

c Using your graph, obtain the values of x and y that give the minimum value of T. Make your method clear and calculate the minimum value of T.

7 A film critic, Verity, must see five films, *A*, *B*, *C*, *D* and *E* over two days. The films are being shown at five special critics' preview times. The bipartite graph shows the times at which each film is showing.

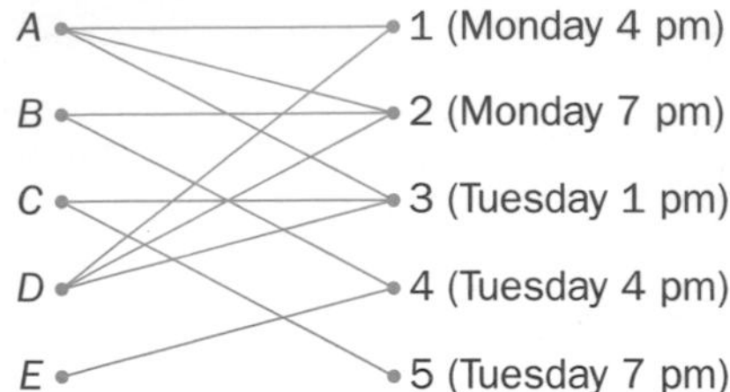

Initially, Verity intends to see Film *A* on Monday at 4 pm, Film *B* on Tuesday at 4 pm, Film *C* on Tuesday at 1 pm and Film *D* on Monday at 7 pm.

a On a copy of the diagram, show this initial matching in a distinctive way.

Using the maximum matching algorithm and the given initial matching,

b find two distinct alternating paths and complete the matchings they give.

Verity's son is keen to see film *D*, but he can only go with his mother to the showing on Monday at 7 pm.

c Explain why it will not be possible for Verity to take her son to this showing and still see all five films herself.

8

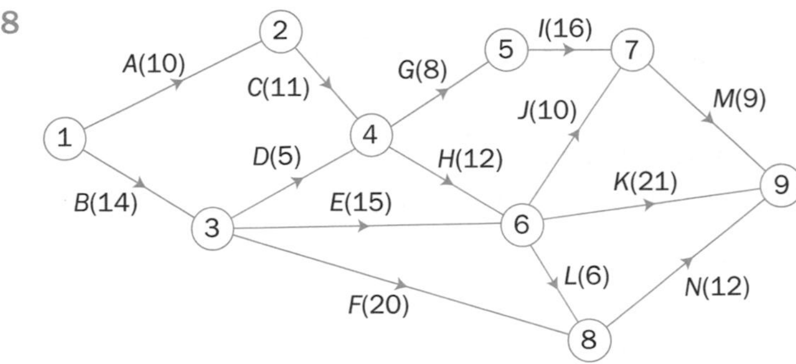

An engineering project is modelled by the activity network shown. The activities are represented by the arcs. The number in brackets on each arc gives the time, in days, to complete the activity. Each activity requires one worker. The project is to be completed in the shortest time.

a Calculate the early time and late time for each event.

b State the critical activities.

c Find the total float on activities *D* and *F*.

d Draw a cascade (Gantt) chart for this project.

The chief engineer visits the project on day 15 and day 25 to check the progress of the work. Given that the project is on schedule,

e which activities **must** be happening on each of these two days?

Answers

Chapter 1

Exercise 1.1

1 1, 1, 2, 3, 5, 8, 13, 21, 34, 55

2 a 1, 4, 9, 16, 25, 36, 49, 64, 81, 100

b Prints the squares of positive integers up to 10

c i 1, 2, 3, 4, 5, 6, 7, 8, 9, 10

ii 1, 3, 6, 10, 15, 21, 28, 36, 45, 55

iii Squares would stop at 81.

3 a i

A	12	6	3	1	
B	7	14	28	56	
C	0	0	28	84	Print 84

ii

A	53	26	13	6	3	1	1	
B	76	152	304	608	1216	2432	2432	
C	0	76	76	380	380	1596	4028	Print 4028

b Multiplies the numbers together

c i Lots of steps before result found

ii Swap the values of *A* and *B* if necessary so that $A < B$

4 a i

A	48	36	12	
B	132	48	36	
R	36	12	0	Print HCF = 12

ii

A	130	78	52	26	
B	78	130	78	52	
R	–	52	26	0	Print HCF = 26

b

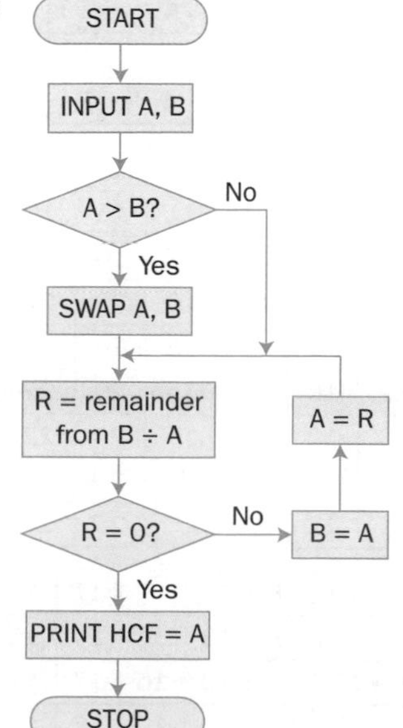

5 a i

A	B	C	D	Test	Print
2	-6	-3	60	>0	Real, distinct roots

ii

A	B	C	D	Test	Print
3	4	5	-44	<0	No real roots

iii

A	B	C	D	Test	Print
9	-6	1	0	=0	Real, repeated root

b

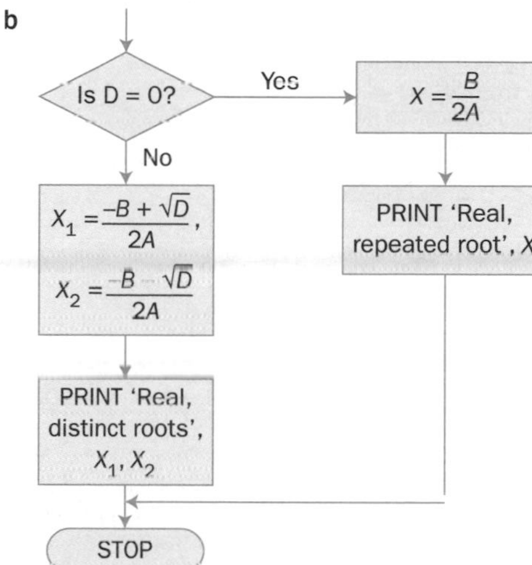

6

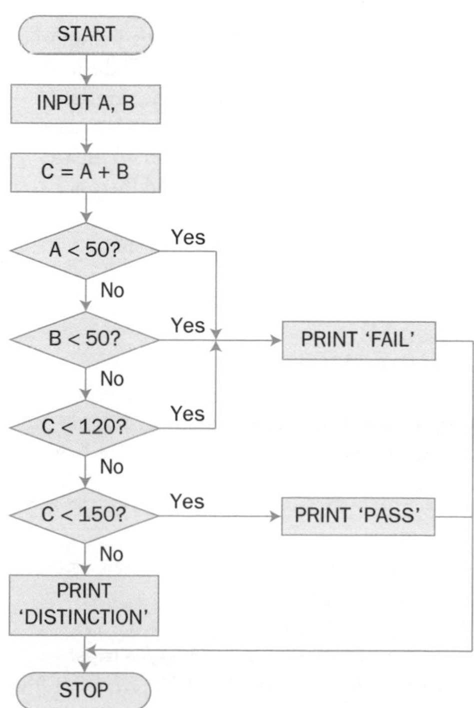

7 **a** **Step 1** Two children cross
Step 2 One child returns
Step 3 One adult crosses
Step 4 One child returns
Step 5 If more adults remain. go to Step 1
Step 6 Stop

b 40 crossings

8

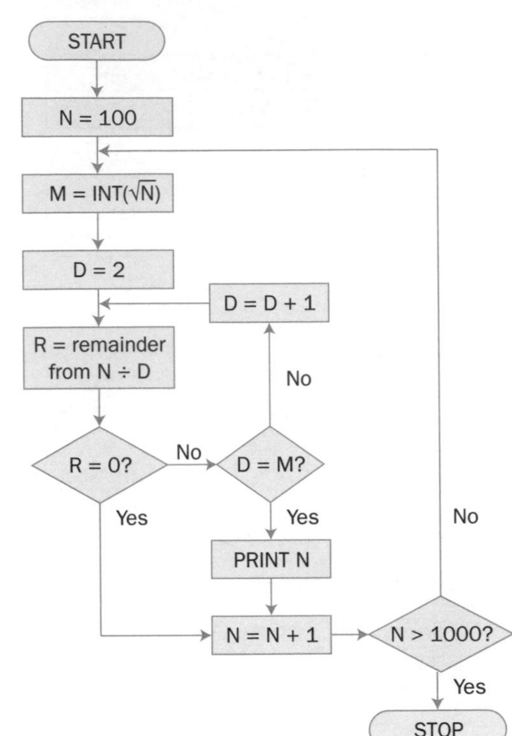

Notes:
(A) INT($\sqrt{N}$) = largest integer $\leqslant \sqrt{N}$
(B) The flowchart shown is one possible approach. It is inefficient because it tries all the integers as divisors, rather than just the primes.

9 Each white counter must move $(n + 1)$ spaces
Each black counter must move $(m + 1)$ spaces
Each jump covers two squares and there are mn jumps
So, moves $= m(n + 1) + n(m + 1) - mn =$
$mn + m + n$

Exercise 1.2

1 **a** At least 4 bins required
b (5, 3, 9), (6, 11), (2, 3, 7, 6), (4, 6), (10)
c (11, 7), (10, 6, 2), (9, 6, 3), (6, 5, 4, 3)

2 **a** e.g. (10, 8, 2), (6, 6, 4, 4), (9, 6, 5), (8, 6, 6), (5, 5, 3, 3, 2, 2)
b (5, 3, 3, 4, 4), (6, 6, 8), (10, 6, 2, 2), (5, 6, 6, 2), (4, 8), (10)
c (10, 10), (8, 8, 4), (6, 6, 6, 2), (6, 6, 5, 3), (5, 4, 4, 3, 2, 2)

3 **a** (3.0, 1.2, 0.5), (3.5, 1.0), (2.5, 1.0, 1.2), (2.5, 0.7, 1.8), (2.0, 2.1), (1.2) = 6 trips
b (3.5, 1.2), (3.0, 2.0), (2.5, 2.5), (2.1, 1.8, 1.0), (1.2, 1.2, 1.0, 0.7, 0.5) = 5 trips
c Total payload = 24.2 tonnes > 4 × 5 tonnes

4 **a** First-fit gives (2, 2, 3, 3), (4, 6), (7), (9)
First-fit decreasing gives (9, 3), (7, 4), (6, 3, 2), (2)
b e.g. (7, 3, 2), (6, 4, 2), (9, 3)

5 e.g. First-fit decreasing gives (12, 3), (11, 4), (9, 6), (9, 6), (8, 6), (5, 5, 2) = success

6 **a** (9, 9, 9, 8), (9, 8, 8, 7), (7, 7, 7, 7, 7), (6, 6, 6, 6, 5, 5), (5, 5, 5, 5, 5, 5, 5) (5, 4, 4, 4, 4, 4, 4, 4), (4)
b e.g. change worker 2 to (9, 8, 8, 6, 4) and worker 4 to (7, 6, 6, 6, 5, 5)

7 **a** (6, 8, 10, 6, 10) (20, 24) (15, 20).
15 minutes left over.
b (24, 20) (20, 15, 10) (15, 10, 8, 6, 6)
c (24, 8, 6, 6) (20, 15, 10) (15, 10, 20)

Exercise 1.3

1 **a**

12	4	16	5	9	2	4
4	12	5	9	2	4	16
4	5	9	2	4	12	16
4	5	2	4	9	12	16
4	2	4	5	9	12	16
2	4	4	5	9	12	16
2	4	4	5	9	12	16

b

12	4	16	5	9	2	4
12	16	5	9	4	4	2
16	12	9	5	4	4	2
16	12	9	5	4	4	2

No swaps in last pass, so stop.

2 **a**

12	4	16	5	9	2	4
4	2	4	5	12	16	9
2	4	4	5	12	9	16
2	4	4	5	9	12	16

b

12	4	16	5	9	2	4
12	16	9	5	4	2	4
16	12	9	5	4	4	2
16	12	9	5	4	4	2

3

22	26	14	20	12	9	11	15	10
22	14	20	12	9	11	15	10	26
14	20	12	9	11	15	10	22	26
14	12	9	11	15	10	20	22	26
12	9	11	14	10	15	20	22	26
9	11	12	10	14	15	20	22	26
9	11	10	12	14	15	20	22	26
9	10	11	12	14	15	20	22	26
9	10	11	12	14	15	20	22	26

4

9	17	6	19	16	13	7	17	12	9
9	6	7	12	9	13	17	19	16	17
6	7	9	12	9	13	16	17	19	17
6	7	9	9	12	13	16	17	17	19
6	7	9	9	12	13	16	17	17	19

5 a

H	T	P	F	C	A	L
H	P	F	C	A	L	T
H	F	C	A	L	P	T
F	C	A	H	L	P	T
C	A	F	H	L	P	T
A	C	F	H	L	P	T
A	C	F	H	L	P	T

b

H	T	P	(F)	C	A	L
C	(A)	F	H	T	(P)	L
A	C	F	H	(L)	P	T
A	C	F	H	L	P	T

6 a

1	5	9	3	11	7	13
1	5	9	3	11	7	13
1	5	9	3	7	11	13
1	5	7	3	9	11	13
1	5	3	7	9	11	13
1	3	5	7	9	11	13
1	3	5	7	9	11	13

Required 21 comparisons and 4 swaps

b

1	5	9	3	11	7	13
1	5	3	9	7	11	13
1	3	5	7	9	11	13
1	3	5	7	9	11	13

No swaps in last pass, so stop.
Required 18 comparisons and 4 swaps.
Bubble sort was more efficient.

7 a 10 comparisons, 10 swaps
b i 45 of each ii 190 of each
iii $\frac{n(n-1)}{2}$ of each.

8 a 1, 3, 5, 7, 6, 4, 2 b 21 comparisons

Exercise 1.4

1 a Garton b Garstang c Lada

2 a Examine Garton, search Nish – Quedgeley
Examine Patel, search Nish
Examine Nish – found.
b Examine Garstang, search Dewsbury – Exeter
Examine Evesham, search Exeter
Examine Exeter, not found
c Examine Lada, search Kia – Fiat
Examine Hyundai, search Jaguar – Kia
Examine Kia, search Jaguar
Examine Jaguar, found.

3 a Examine record 11 (Laurence), search records 12 – 20
Examine record 16 (Quentin), search records 12 – 15
Examine record 14 (Omar), found.
b Examine record 11 (Laurence), search records 1 – 10
Examine record 6 (Edith), search records 7 – 10
Examine record 9 (Ingrid), search records 7 – 8
Examine record 8 (Gerald), search record 7
Examine record 7 (Floella), found.
c Examine record 11 (Laurence), search records 12 – 20
Examine record 16 (Quentin), search records 12 – 15
Examine record 14 (Omar), search records 12 – 13
Examine record 13 (Mabel), not found.

4 a Examine record 16, search records 1 – 15
Examine record 8, search records 9 – 15
Examine record 12, search records 13 – 15
Examine record 14, search record 13
Examine record 13, found I. R. Smith.
b Examine record 16, search records 17 – 30
Examine record 24, search records 25 – 30
Examine record 28, search records 25 – 27
Examine record 26, search record 27
Examine record 27, found S. Elliott

5 Examine interview 21, search interviews 22 – 40
Examine interview 31, search interviews 22 – 30
Examine interview 26, search interviews 27 – 30
Examine interview 29, search interview 30
Examine interview 30, not found.

6

Guess	Response	New search area
K8	Right, up	L9 – T15
P12	Left, down	L9 – O11
N10	Left, down	L9 – M9
M9	Found	

7 a i 2 ii 3 iii 3 iv 4 v 4
b Number of records is m, where $2^{m-1} \leqslant n < 2^m$.
If you have met logarithms, this is
$m = \text{INT}(\log_2 n) + 1$
c i 3 ii 4 iii 7 iv 8 v 15.
For n items, examine n records.
d i For n records, mean $= \frac{1}{2}(n+1)$
ii For n records and with m as defined in part **b**,

$$\text{mean} = \frac{m(n+1) - 2^m + 1}{n}$$

Review 1

1 a 1, 8, 27, 64, 125, 216, 343, 512
b Prints the cubes of positive integers

2 a i

A	12	6	3	
C	0	1	2	No

ii

A	8	4	2	1	
C	0	1	2	3	Yes, 3

iii

A	30	15	
C	0	1	No

iv

A	32	16	8	4	2	1	
C	0	1	2	3	4	5	Yes, 5

b It identifies if A is of the form 2^C, and prints the value of C.

3 a 1, 1, 2, 3, 5, 8, 13, 21, 34, 55

b

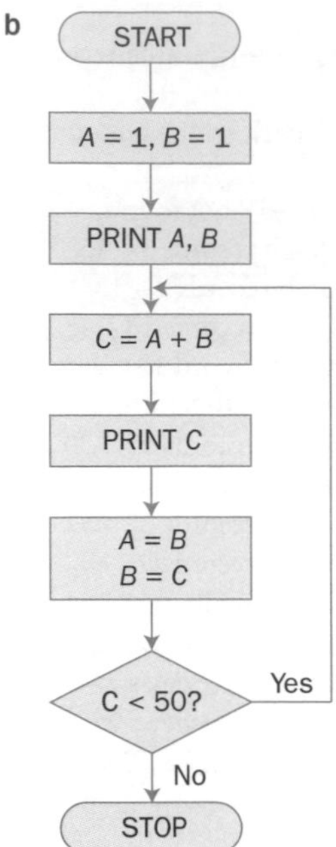

4

C8	A5	D15	S7	M13	K9	T12	U6	P9	F10
C8	A5	S7	U6	P9	K9	D15	M13	T12	F10
A5	U6	S7	C8	P9	K9	F10	T12	D15	M13
A5	U6	S7	C8	P9	K9	F10	T12	M13	D15

5 a (4, 6, 5, 12, 5, 5), (14, 6, 4, 4, 10), (14, 16, 5, 5), (6, 8, 13, 6, 4) + 4 left over.

b (12, 14, 14), (16, 10, 13), (4, 6, 5, 5, 5, 6, 4, 4), (5, 6, 5, 8, 6, 4, 4)

6 a At least 5 trips required

b (9, 6), (9, 6), (8, 5), (8, 5), (4, 4, 4, 3) + 3 left over.

c e.g. (9, 6), (9, 6), (8, 4, 3), (8, 4, 3), (5, 5, 4)

7 Using first-fit decreasing: (2, 2), (2, 1.5, 0.5), (1.2, 1.2, 0.7, 0.7), (0.7, 0.6, 0.6, 0.4, 0.4, 0.4, 0.3, 0.3), waste = 0.5 m

8

12	4	9	15	11	12	8	2	7
4	9	12	11	12	8	2	7	15
4	9	11	12	8	2	7	12	15
4	9	11	8	2	7	12	12	15
4	9	8	2	7	11	12	12	15
4	8	2	7	9	11	12	12	15
4	2	7	8	9	11	12	12	15
2	4	7	8	9	11	12	12	15
2	4	7	8	9	11	12	12	15

9

12	4	9	15	11	12	8	2	7
4	9	8	2	7	11	12	15	12
4	2	7	8	9	11	12	12	15
2	4	7	8	9	11	12	12	15
2	4	7	8	9	11	12	12	15

10

Hu	Be	Ca	Le	Je	Af	Br	Bu	Wi	Bl
Be	Ca	Hu	Je	Af	Br	Bu	Le	Bl	Wi
Be	Ca	Hu	Af	Br	Bu	Je	Bl	Le	Wi
Be	Ca	Af	Br	Bu	Hu	Bl	Je	Le	Wi
Be	Af	Br	Bu	Ca	Bl	Hu	Je	Le	Wi
Af	Be	Br	Bu	Bl	Ca	Hu	Je	Le	Wi
Af	Be	Br	Bl	Bu	Ca	Hu	Je	Le	Wi
Af	Be	Bl	Br	Bu	Ca	Hu	Je	Le	Wi
Af	Be	Bl	Br	Bu	Ca	Hu	Je	Le	Wi

No swap in last pass, so stop.

11 a

125	86	315	97	130	266	45	120	284	90
86	125	315	97	130	266	45	120	284	90
86	125	315	97	130	266	45	120	284	90
86	97	125	315	130	266	45	120	284	90
86	97	125	130	315	266	45	120	284	90
86	97	125	130	266	315	45	120	284	90
45	86	97	125	130	266	315	120	284	90
45	86	97	120	125	130	266	315	284	90
45	86	97	120	125	130	266	284	315	90
45	86	90	97	120	125	130	266	284	315

30 comparisons, 23 swaps

b

125	86	315	97	130	266	45	120	284	90
86	125	97	130	266	45	120	284	90	315
86	97	125	130	45	120	266	90	284	315
86	97	125	45	120	130	90	266	284	315
86	97	45	120	125	90	130	266	284	315
86	45	97	120	90	125	130	266	284	315
45	86	97	90	120	125	130	266	284	315
45	86	90	97	120	125	130	266	284	315
45	86	90	97	120	125	130	266	284	315

44 comparisons, 23 swaps

12 a

55	80	25	84	25	34	17	75	3	5
80	55	84	25	34	25	75	17	5	3
80	84	55	34	25	75	25	17	5	3
84	80	55	34	75	25	25	17	5	3
84	80	55	75	34	25	25	17	5	3
84	80	75	55	34	25	25	17	5	3
84	80	75	55	34	25	25	17	5	3

No swaps on last pass, so stop.

b At least 5 bins needed.

c (84, 5, 3), (80, 17), (75, 25), (55, 34), (25)

13 a Examine record 9 (Mountjoy), search records 10 – 16
Examine record 13 (Trueman), search records 10 – 12
Examine record 11 (Simpson), found.

b Examine record 9 (Mountjoy), search records 1 – 8
Examine record 5 (Harris), search records 1 – 4
Examine record 3 (Faruq), search record 4
Examine record 4 (Harman), found.

c Examine record 9 (Mountjoy), search records 10 – 16
Examine record 13 (Trueman), search records 10 –12
Examine record 11 (Simpson), search record 12
Examine record 12 (Thompson), not found.

14 a i Examine record 13 (Lapwing), search records 1 – 12
Examine record 7 (Goldfinch), search records 8 – 12
Examine record 10 (Heron), found.
(3 records examined)

ii Examine record 13 (Lapwing), search records 14 – 25
Examine record 20 (Starling), search records 14 – 19
Examine record 17 (Robin), search records 14 – 16
Examine record 15 (Pheasant), search record 14
Examine record 14 (Nuthatch), found.
(5 records examined)

iii Examine record 13 (Lapwing), search records 14 – 25
Examine record 20 (Starling), search records 21 – 25
Examine record 23 (Widgeon), search records 21 – 22
Examine record 22 (Whitethroat), search record 21
Examine record 21 (Swallow), not found.
(5 records examined)

b i 10 ii 14 iii 25

15 a The list is not in alphabetical order.

b Sort using either bubble or quick sort algorithm.
Bubble:

G	N	M	Y	L	B	C	E	S	P
G	M	N	L	B	C	E	S	P	Y
G	M	L	B	C	E	N	P	S	Y
G	L	B	C	E	M	N	P	S	Y
G	B	C	E	L	M	N	P	S	Y
B	C	E	G	L	M	N	P	S	Y
B	C	E	G	L	M	N	P	S	Y

No swaps on last pass, so stop.
Quick sort:

G	N	M	Y	L	(B)	C	E	S	P
B	G	N	M	Y	(L)	C	E	S	P
B	G	(C)	E	L	N	M	(Y)	S	P
B	C	G	(E)	L	N	M	(S)	P	Y
B	C	E	G	L	N	(M)	P	S	Y
B	C	E	G	L	M	N	(P)	S	Y
B	C	E	G	L	M	N	P	S	Y

c Examine record 6 (Manchester), search records 7 – 10
Examine record 9 (Southampton), search records 7 – 8
Examine record 8 (Plymouth), search record 7
Examine record 7 (Newcastle), found.

Chapter 2

Exercise 2.1

1 (these are examples – others may be possible)

a **b**

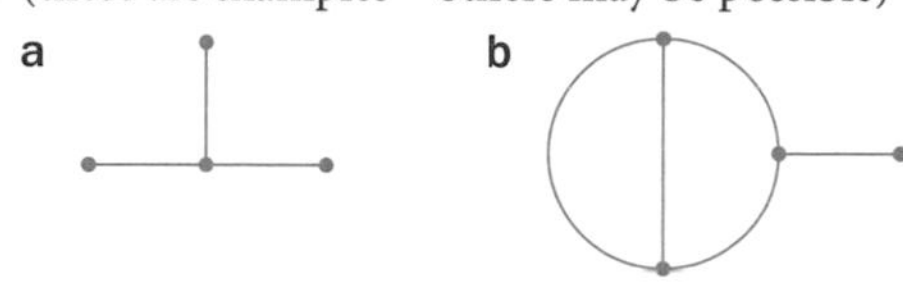

c

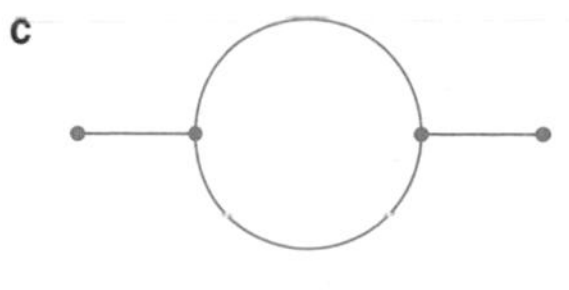

d

2 a Graph i **b** Graph iv **c** Graph ii

3 (your layout may be different)

a **b**

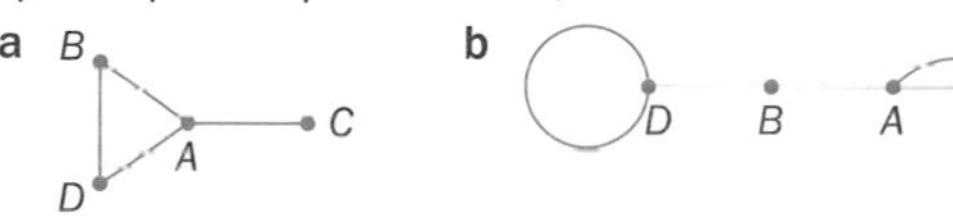

4

	A	B	C	D	E
A	0	1	1	1	1
B	1	0	1	0	1
C	1	1	0	2	0
D	1	0	2	0	1
E	1	1	0	1	0

5 a {*A*, *B*, *C*, *D*, *E*, *F*}, {*AB*, *AC*, *AD*, *BC*, *BE*, *CF*, *DE*, *DF*, *EF*}

b {*A*, *B*, *C*, *D*, *E*}, {*AB*, *AD*, *AE*, *BC*, *BE*, *CD*, *CD*, *CE*}

6

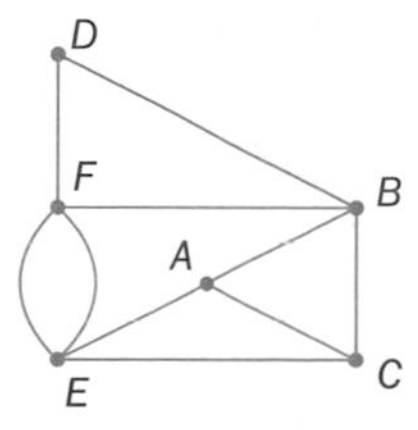

7 Graph **d** is not a subgraph.

a {*A*, *B*, *C*}, {*AB*, *AC*, *BC*}

b {*A*, *B*, *D*, *E*}, {*AB*, *AD*, *BE*, *DE*}

c {*A*, *B*, *C*, *D*, *E*, *F*}, {*AD*, *BE*, *CF*, *DE*, *EF*}

e {*A*, *B*, *D*, *E*, *F*}, {*AB*, *AD*, *BE*, *DF*, *EF*}

f {*A*, *B*, *C*, *D*, *E*, *F*}, {*AB*, *AD*, *BC*, *CF*, *DE*, *EF*}

8 a **b**

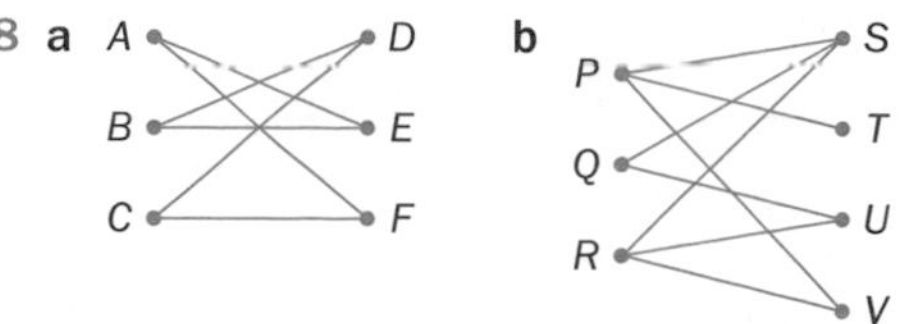

9 a

	A	B	C	D	E	F
A	0	1	0	1	0	1
B	1	0	1	0	0	0
C	0	1	0	1	0	1
D	1	0	1	0	1	0
E	0	0	0	1	0	1
F	1	0	1	0	1	0

b

	A	C	E	B	D	F
A	0	0	0	1	1	1
C	0	0	0	1	1	1
E	0	0	0	0	1	1
B	1	1	0	0	0	0
D	1	1	1	0	0	0
F	1	1	1	0	0	0

{A, C, E} and {B, D, F}

c

	B	D	F
A	1	1	1
C	1	1	1
E	0	1	1

10 a i 2 ii 3 iii 4 iv $(n-1)$

b i 3 ii 6 iii 10 iv $\frac{1}{2}n(n-1)$

c mn

d If v is even, max edges $= \frac{v^2}{4}$.

If v is odd, max edges $= \frac{(v^2-1)}{4}$

11 a

	A	B	C	D	E	F
A	–	13	5	9	–	17
B	13	–	6	–	11	10
C	5	6	–	–	6	–
D	9	–	–	–	12	28
E	–	11	6	12	–	–
F	17	10	–	28	–	–

b e.g. $AB > (AC + BC)$

12

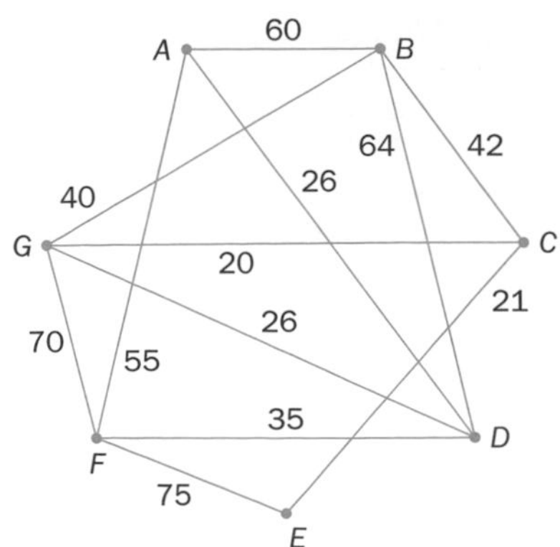

Exercise 2.2

1 a i {A, B, C, D, E}, {AD, AE, BA, BC, CD, CE, DC, DC, DE, EB, EC}

ii

		To				
		A	B	C	D	E
From	A	0	0	0	1	1
	B	1	0	1	0	0
	C	0	0	0	1	1
	D	0	0	2	0	1
	E	0	1	1	0	0

b i {P, Q, R, S, T}, {PP, PQ, PT, QR, QS, RQ, RT, SP, ST, SR}

ii

		To				
		P	Q	R	S	T
From	P	1	1	0	0	1
	Q	0	0	1	1	0
	R	0	1	0	0	1
	S	1	0	1	0	1
	T	0	0	0	0	0

2

3

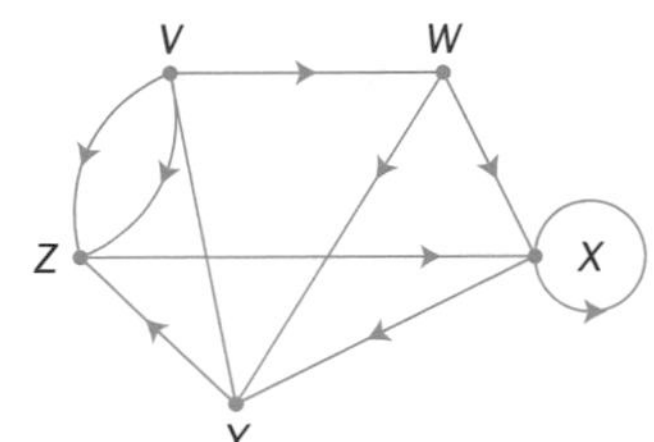

4

	A	B	C	D	E	F
A	–	7	4	–	–	–
B	–	–	–	–	–	9
C	–	2	–	–	4	–
D	8	–	–	–	3	7
E	–	–	–	–	–	6
F	–	–	–	–	–	–

5

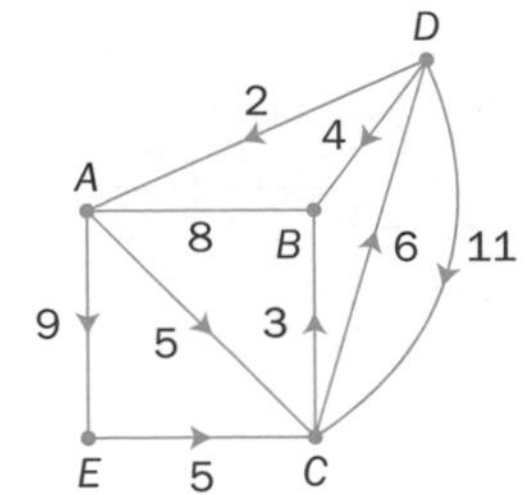

Exercise 2.3

1 D A K J — 1 2 3 4

2

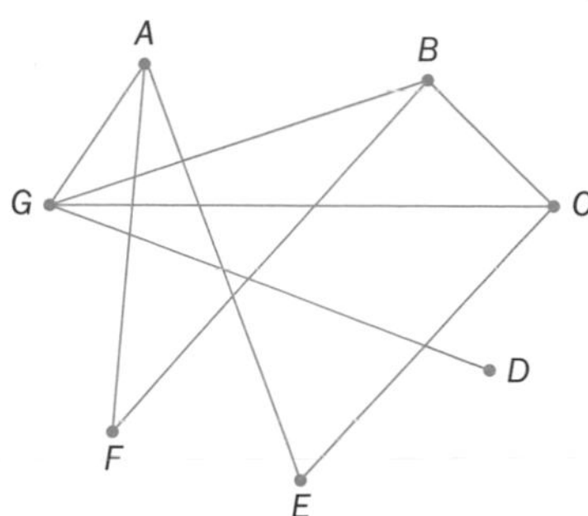

3 B C D E F — bricklaying, carpentry, painting, plumbing, glazing, electrical

4

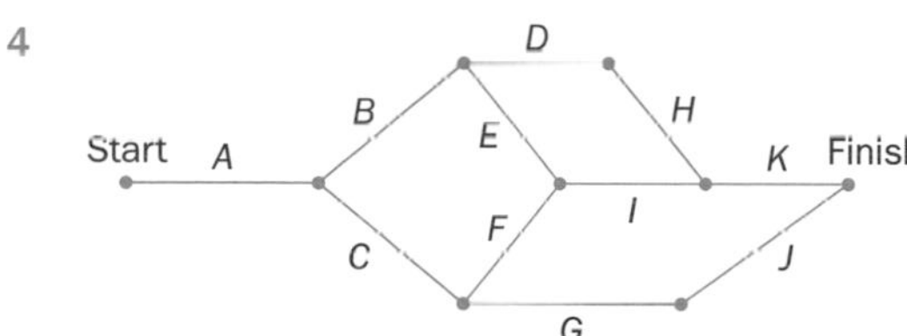

5

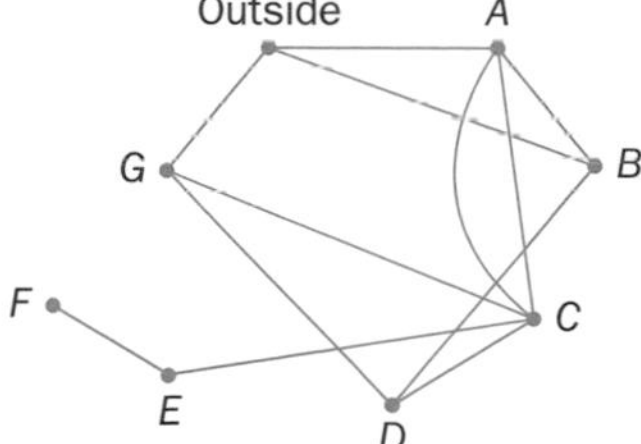

6 **a** slug, greenfly **b** spider, slug
c i lacewing, frog, hoverfly ii swallow

7

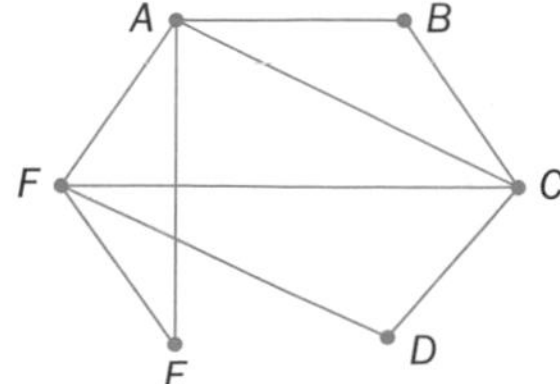

Exercise 2.4

1 **a** i *ABC* ii *ADFC* iii *ADEFC*
iv *ABEDFC* (these are just examples)
b i *BACB* ii *BADEB* iii *BADFEB*
iv *BADEFCB* (these are just examples)
c It visits *C* twice

2 Graphs **a**, **d** and **e**

3 **a** 3 **b** 6

Exercise 2.5

1 **a** neither **b** Eulerian
c semi-Eulerian **d** Eulerian
e semi-Eulerian **f** neither

2 **a**

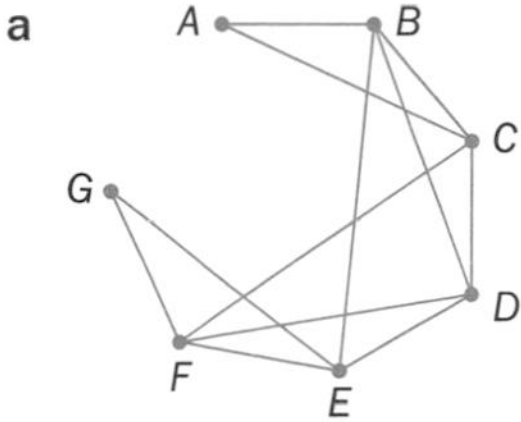

b The degrees of vertices – *A*(2), *B*(4), *C*(4), *D*(4), *E*(4), *F*(4), *G*(2) – are all even
c e.g. *ABCDEFGEBDFCA*
d The sum of each row (or column) in the table gives the degree of that vertex

3 K_n is traversable if and only if n is odd.

4 **a**

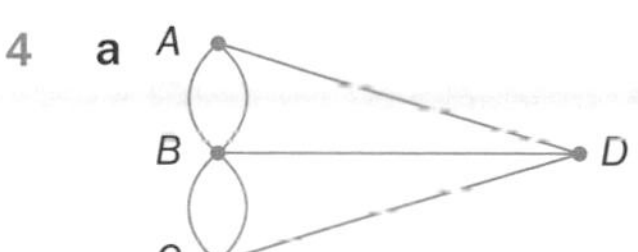

b All odd nodes, so graph is not traversable.

Review 2

1 **a** *A*(3), *B*(3), *C*(4), *D*(4), *E*(3), *F*(3)
b 10 edges.
Sum of degrees = 2 × number of edges.
c {*A, B, C, D, E, F*}, {*AB, AC, AD, BC, BF, CD, CE, DE, DF, EF*}

d

	A	B	C	D	E	F
A	0	1	1	1	0	0
B	1	0	1	0	0	1
C	1	1	0	1	1	0
D	1	0	1	0	1	1
E	0	0	1	1	0	1
F	0	1	0	1	1	0

Sums of columns give degrees of vertices

2 **a**

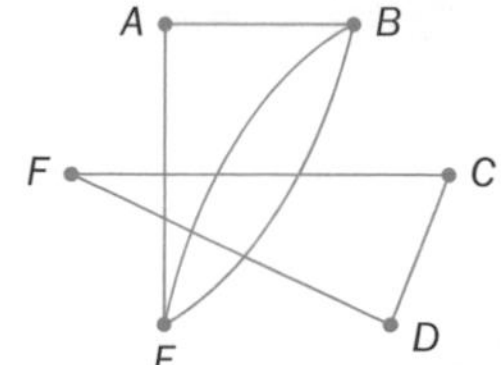

b Not connected – can't travel e.g. from *A* to *C*
c Not simple – *B* and *E* connected by 2 arcs

D1

3 a

	A	B	C	D	E
A	0	1	1	1	1
B	1	0	1	0	1
C	1	1	0	1	1
D	1	0	1	0	0
E	1	1	1	0	0

b 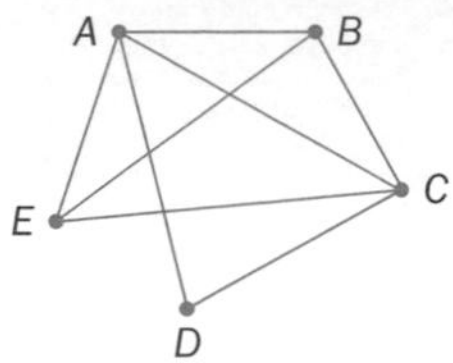

c 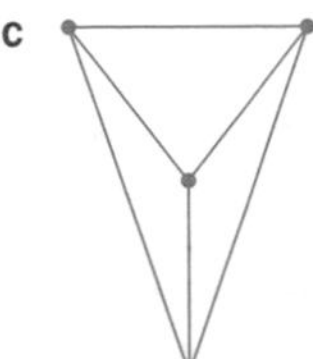

d Yes, $\{A, B, C, E\}$

4 a Graphs vii, viii b Graph iii
c (i, vi), (ii, v), (iv, ix, x), (vii, viii)

5 a

b 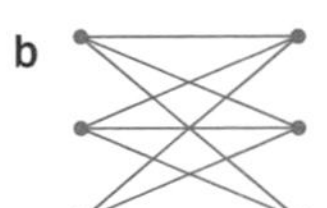

c Yes, $\{A, C, D, F, G, H\}$

6 a

b No. Lee is the only one who plays drums and bass, and can't do both.

7 a

	P	E	F	G	A
P	–	7	22	–	–
E	7	–	–	8	15
F	22	–	–	20	–
G	–	8	20	–	4
A	–	15	–	4	–

b No, e.g. $EG + GA < EA$

8 a $\{A, B, C, D, E\}$, $\{AB, AD, AD, AE, BC, CA, CD, DA, DC, EB, EC, ED\}$

b

		To				
		A	B	C	D	E
From	A	0	1	0	2	1
	B	0	0	1	0	0
	C	1	0	0	1	0
	D	1	0	1	0	0
	E	0	1	1	1	0

9

10

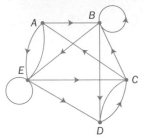

		To				
		A	B	C	D	E
From	A	–	12	20	–	–
	B	–	–	8	–	18
	C	20	8	–	9	10
	D	14	–	–	–	–
	E	–	–	–	15	–

11 a No, repeat of B and C
b Yes, a closed path
c False, all degrees are even

12 e.g.

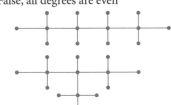

10 hydrogen in each case.

13 a 2 odd nodes (B and D)
b BD
c e.g. $ABCDEFGEADBFCA$

14 a 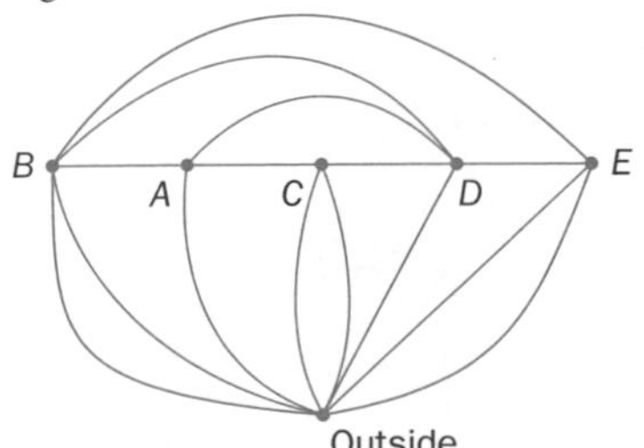

b 4 odd nodes A, B, D and Outside.
c i 2 doors ii A-B and D-Outside.

15 a 5 b 15 c 9 d 6
e Two of the connected nodes would be in the same set.

16 a

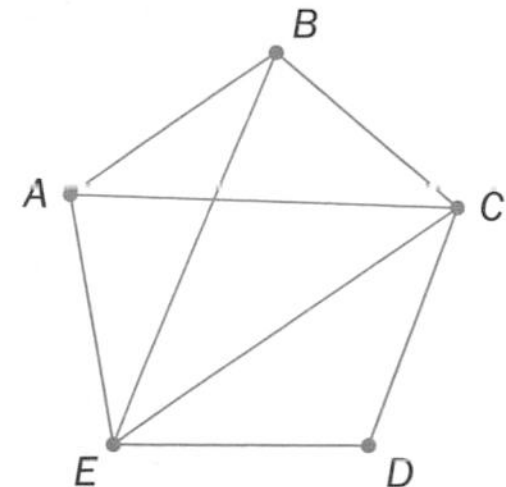

b i *ABCD*, *ABED*, *ABCED*, *ABECD*, *ACD*, *ACED*, *ACBED*, *AED*, *AECD*, *AEBCD*

ii *ABCDEA*, *ABEDCA*

Chapter 3

Exercise 3.1

1 **a** *CD*, *AB*, *BD*, (not *AD* or *AC*), *BE*. Total 22

b *BC*, *DE*, *AB*, *FE*, (*GI* or *HI*), (*HI* or *GI*), (*AH* or *BH*), *IF*. Total 307

2 **a** *AB*, *BD*, *DC*, *BE*

b *AB*, *BC*, *AH* (or *BH*), *HI*, *IG*, *IF*, *FE*, *ED*

3 *AB*, *BD*, (not *AD*), *AC* or *AB*, *BD*, (not *AD*), *BC*. Both total 7.

4 **a** Two solutions – total 20 in each case *AF*, *FG*, *FD*, (*DB* or *DC*), (*DC* or *DB*), *GE* or *AF*, *FG*, *GE*, *FD*, (*DB* or *DC*), (*DC* or *DB*)

b *AH*, (*HG* or *HI*), (*HI* or *HG*), *IC*, *AB*, *CD*, *DE*, *EF*. Total 178.

5 **a**

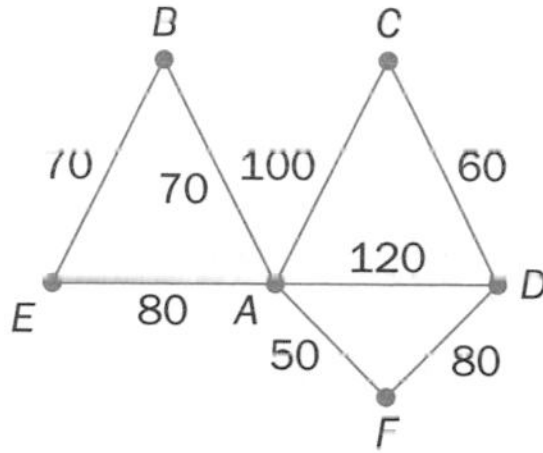

b *AF*, *CD*, (*AB* or *BE*), (*BE* or *AB*), (not *AE*), *DF* Total 330 m

6 *AG*, *AB*, *BC*, *CH*, *CD*, *DI*, *IF*, *FE*. 10 tonnes

7 **a**

	A	B	C	D	E
A	–	5	11	8	6
B	5	–	6	13	4
C	11	6	–	8	8
D	8	13	8	–	10
E	6	4	8	10	–

b

c *AB*, *BE*, *BC*, (*AD* or *CD*). Total 23

d The same stretch of road is involved in two distances.

e

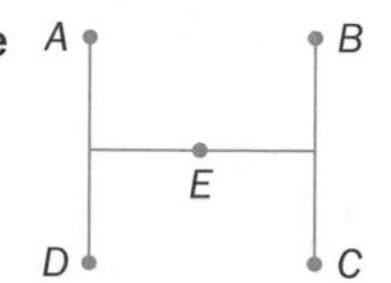

Total 21

Exercise 3.2

1 *AE*, *AB*, *BD*, *CD*, chosen in that order. Total weight 17

2 *FA*, *AB*, *BE*, *FD*, *DC*. Total 330 m

3 **a** *PQ*, *QR*, *QT*, *ST*, *TU* or *PQ*, *QR*, *RS*, *ST*, *TU*. Total 14

b

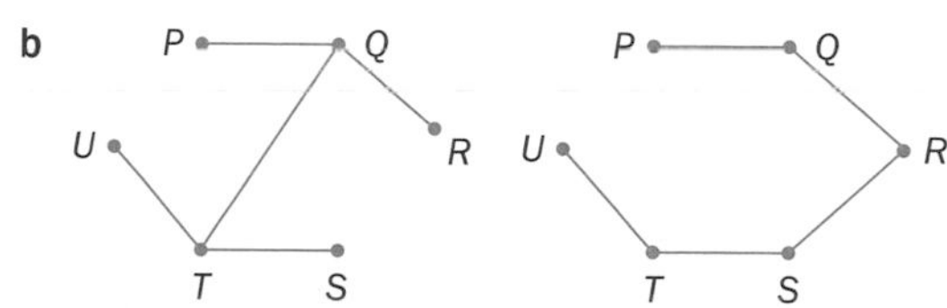

4 *AC*, *CH*, *AB*, *BD*, *DG*, *GE*, *EF*. Total 216

5 **a** *GE* (30), *ES* (22), *SI* (20), *IF* (22), *SP* (23)

b Cheapest sequence is

Total cost £1170

c Cost £1720. May be better because

i it would take less time if all translations were made in parallel

ii slight changes of meaning might build up through a succession of translations.

6 **a**

	O	A	B	C	D	E	F
O	–	4	7	8.5	6.5	4.5	3
A	4	–	3	4.5	4.5	5.5	3
B	7	3	–	2.5	3.5	4.5	4
C	8.5	4.5	2.5	–	2	4	5.5
D	6.5	4.5	3.5	2	–	2	3.5
E	4.5	5.5	4.5	4	2	–	2.5
F	3	3	4	5.5	3.5	2.5	–

b *OF*, *FE*, *ED*, *DC*, *CB*, (*AB* or *AF*). Total 15 m.

c Yes, improvement is possible. The author knows of no general algorithm for this problem, but a makeshift strategy gave the following possible solution, length 13.5 m.

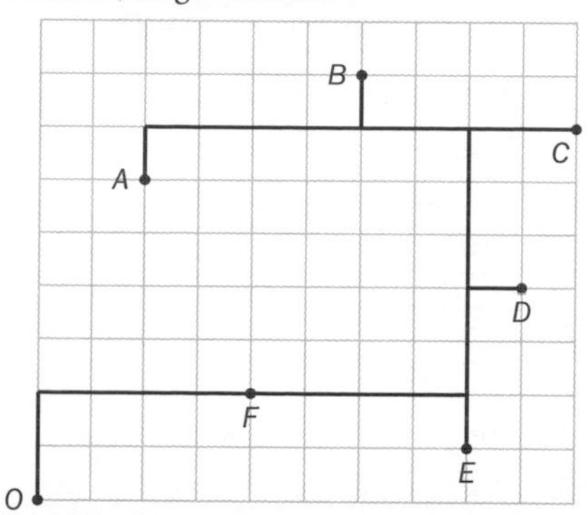

Review 3

1 *DG*, *FG* or *BC*, *BC* or *FG*, *AD* or *EC*, *EC* or *AD*, (not *DF*), (not *BE*), *GH*, *AB*. Total 36

2 *AC*, *CE*, *EB*, *CG*, *AD*, *BF* or *GF*. Total 96

3 *AD*, *DG*, *AB*, *BC*, *CF*, *DE*. Total 47

4 **a** Difficult to tell if a cycle is being formed. Prim's avoids this possibility.

b *AB*, *BC*, *BD* or *CD*, *CE*, *EF*.

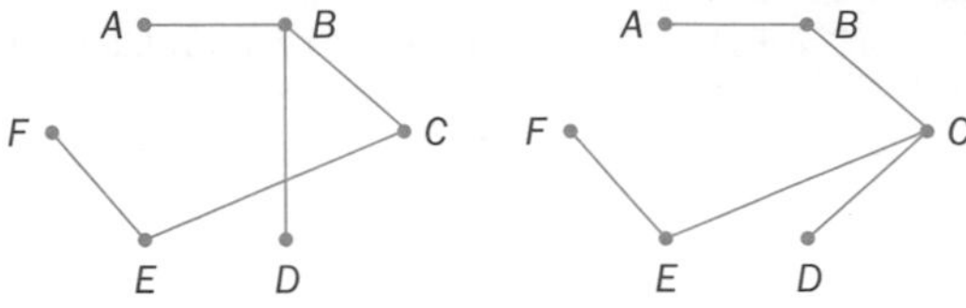

c i 32 km

ii For first tree, bus station at *C*, longest journey 23 km. For second tree, bus station at *E*, longest journey 21 km

5 **a** *EG* or *BG*, *BG* or *EG*, (not *BE*), *AC* or *BD*, *BD* or *AC*, *CD* or *DF*, *DF* or *CD* Total 24 km

b *CD*

6 **a** *AF*, *EF*, *DE*, *AB*, *CF*. Total 16 minutes

b

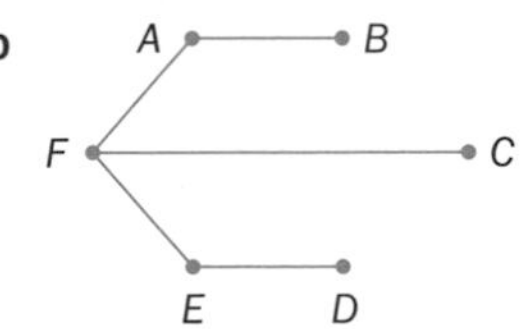

c 11 minutes

7 **a** i *FH*, *AD*, *DE*, *CE*, *CB* or *EG*, *EG* or *CB*, *CF*, *HI*, *IJ*

ii *AD*, *DE*, *CE*, *CB* or *EG*, *EG* or *CB*, *CF*, *FH*, *HI*, *IJ*

b Use Kruskal with *AB*, *FI* already chosen

Chapter 4

Exercise 4.1

1 The labels on the vertices should be:

S	1	0	A	3	6	B	4	9	C	2	5	D	6	13
			6			9			5			~~15~~ 13		

E	5	12	F	7	16	G	8	19	T	9	26
~~13~~ 12			~~19~~ 16			~~26~~ 19			~~29~~ ~~28~~ 26		

The shortest route is *SBEFGT* = 26

2 The labels on the vertices should be:

S	1	0	A	3	6	B	2	3	C	4	8	D	6	15
			6			3			8			~~16~~ 15		

E	5	11	T	7	19
11			19		

Shortest routes *SBCDT* or *SBCT*, total 19.

3 **a, b**

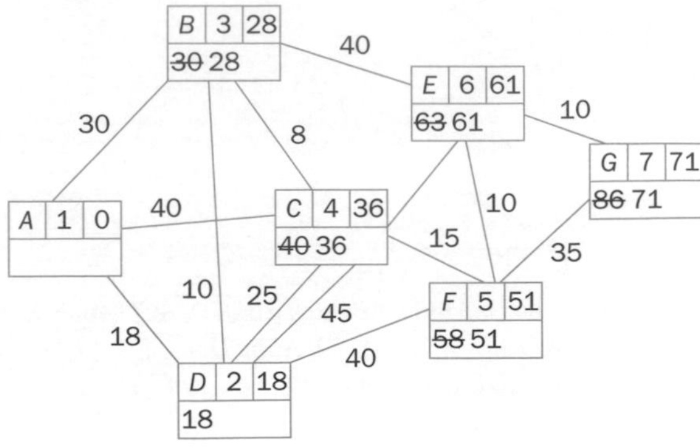

Shortest route *ABCFEG*, total 71 minutes.

c The table doesn't allow for waiting times.

4 *A* to *G*: *ACEDG*, total 25. *G* to *A*: *GFCEA*, total 21.

5 Warehouse *B*. Route *BHJL*, total 25 km

6 **a** *ABCEF*, total £200.

b *ADF*, total £240.

7 18 tonnes, route *ABEGH* or *ACBEGH*.

Review 4

1 The labels on the vertices should be:

S	1	0	A	2	4	B	4	7	C	3	5	D	6	16
			4			~~8~~ 7			5			16		

E	5	11	F	7	21	T	8	23
~~12~~ 11			21			23		

The shortest route is *SABET*, 23 km.

2 **a** The labels on the vertices should be:

A	1	0	B	5	7	C	6	11	D	8	13	E	4	6
			~~8~~ 7			11			~~21~~ 13			6		

F	2	2	G	3	6	H	7	11
2			6			~~16~~ 11		

The shortest route is *AFGHD*, 13 minutes.

b Route unaltered, takes 16 minutes

c *AFGBCD*, 16 minutes

3 **a** The labels on the vertices should be:

A	1	0	B	5	17	C	3	18	D	2	15	E	6	32
			17			18			15			~~35~~ ~~33~~ 32		

F	4	28	G	9	38	H	7	44	I	8	41	J	10	53
~~29~~ 28			44 38			~~45~~ 44			44 41			53		

ACFEGJ, 53 km

b Work back from *J*. If *Y* is on the route and (label *Y* – label *X*) = weight *XY*, then *X* is on the route.

c *ACEGJ* or *ADFEGJ*, both 54 km.

4 **a** The labels on the vertices should be:

S	1	0	A	2	3	B	3	5	C	4	7	D	6	10	E	5	9
			3			5			~~8~~ 7			10			9		

F	7	12	G	8	12	H	9	15	I	10	16	T	11	24
~~13~~ 12			12			15			16			24		

SAEGT, £24

b *SBCFEGT*, £31

5 a The labels on the vertices should be:

P	1	0	*Q*	3	2	*R*	2	1	*S*	4	2	*T*	5	5
			2			1			~~6~~2			~~9~~5		

U	6	6	*V*	8	9	*W*	7	8
~~10~~6			~~10~~9			~~12~~8		

PRSUW or *PQTUW*, both 8 lights.

b *PQTW* or *PQTUVW* or *PRSUVW*, all 12 lights.

6 a The labels on the vertices should be:

S	1	0	*A*	3	16	*B*	2	12	*C*	6	8	*D*	5	22
			16			12			8			~~23~~22		

E	4	30	*F*	7	17	*T*	9	37
~~35~~30			17			~~43~~ ~~40~~37		

SCFT, 37 km.

Route found by working back from *T*. If *Y* is on the route and (label *Y* – label *X*) = weight *XY*, then *X* is on the route.

b *SCFET*, 38 km (by Dijkstra, shortest route from *S* to *E* is *SCFE* = 30)

Chapter 5

Exercise 5.1

1 a Repeat *AC* and *CD*.

b Total length 149. Possible route *ABCDEBDCAECA*.

2 a i Possible pairings *AB*/*CD* =14, *AD*/*BC* = 18, *AC*/*BD* = 18.

ii Repeat *AE*, *EB*, *CD*. Total length 70.

b i Possible pairings *AB*/*CF* = 28, *AC*/*BF* = 29, *AF*/*BC* = 29.

ii Repeat *AB*, *CD*, *DG*, *GF*. Total length 163.

3 a Repeat *AJI* (or *AKI*) and *HLF* (or *HGF*). Total distance 4040 m.

b Only need to repeat edge *AF* giving 3980 m.

c Enter at *A* and leave at *F* (or vice versa); 3380 m.

4 a

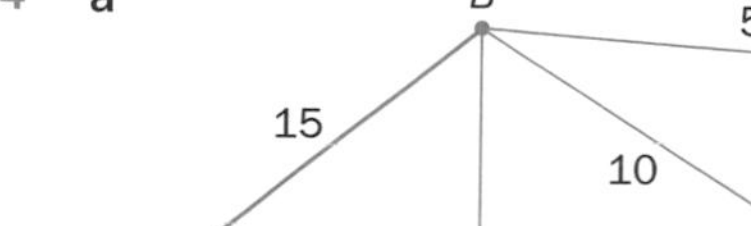
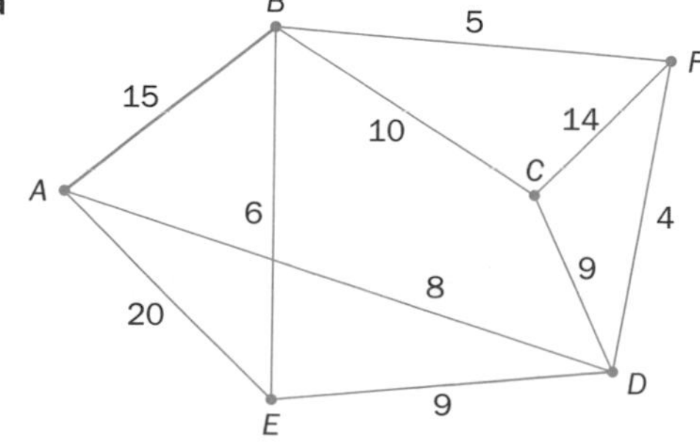

b Repeat *AD*, *DC*, *EB*, *BF* or *AD*, *DF*, *EB*, *BC*. Total distance 128 km.

5 a Repeat e.g. *AB*, *GC*, *CD* and *EF* (or other equivalent edges).

b A possible route is *AHCIEFEDCDIFCGHBAGCBA*. Total distance = 880 m.

6 a 1800 cm, repeating *BJ*, *GI*, *DE* or equivalent.

b 1640 cm. *B* and *G* as start and end. *IJ* and *DE* doubled.

c Min 3 cables e.g. *ILKJABCDMNE*, *JIHGFED*, *BG*

Review 5

1 a Each arc is effectively two arcs, so all nodes are even. Network is Eulerian and distance = 2 × sum of weights.

b Repeat *AB* and *DE*, total distance 108 km.

2 a Distance 4050 m. Repeat *AF*, *FG*, *GC*, *DE*.

b If she does not, then *B* and *G* are odd nodes and other (longer) routes will need repeating.

3 a i 2160 m ii *BD* and *EC*

b i *C*, *E* ii 1950 m iii *BD*

4 a Repeat *AC* and *DF*. Total distance 94 km.

b *D* and *F* giving distance 85 km. Only need to repeat *AC*.

5 a The sum of the degrees of the nodes is even for any graph. The number of odd nodes must therefore be even.

b Repeat *BC*, so total length = $(10.5x - 26)$. This gives $x = 12$

Revision 1

1 a

90	50	55	40	20	35	30	25	45
90	55	50	40	35	30	25	45	20
90	55	50	40	35	30	45	25	20
90	55	50	40	35	45	30	25	20
90	55	50	40	45	35	30	25	20
90	55	50	45	40	35	30	25	20
90	55	50	45	40	35	30	25	20

No swap on last pass, so stop.

b Total 475 mins. At least 4 tapes needed.

c (90, 30), (55, 50), (45, 40, 35), (35, 30, 25, 20), (20)

d e.g. (90, 30), (50, 40, 30), (55, 25, 20, 20), (45, 35, 35)

2 a

45	56	37	79	46	18	90	81	51
45	37	18	46	56	79	90	81	51

b

45	56	37	79	46	18	90	81	51
56	45	79	46	37	90	81	51	18

c Examine record 6 (44), search records 7 – 11
Examine record 9 (71), search records 10 – 11
Examine record 11 (94), search record 10
Examine record 10 (73), found.

3 a Devising route for lorries gritting roads.

b Odd nodes *A*, *B*, *D*, *F*. Best pairing (*AB* + *DF*) = 41, route repeats *AE*, *EB*, *DF*

c Not unique – pairing (*AD* + *BF*) = 41, route repeats *AD* and *BF*.

d 299

e *A* and *B*, only need to repeat *DF*.

4 a

a	b	c	Integers?	Output list	$a = b$?
90	2	45	Y	2	N
45	2	22.5	N		
45	3	15	Y	3	N
15	2	7.5	N		
15	3	5	Y	3	N
5	2	2.5	N		
5	3	1.67	N		
5	5	1	Y	5	Y

b Prime factorisation of a. c $c = 1$.

5 a i

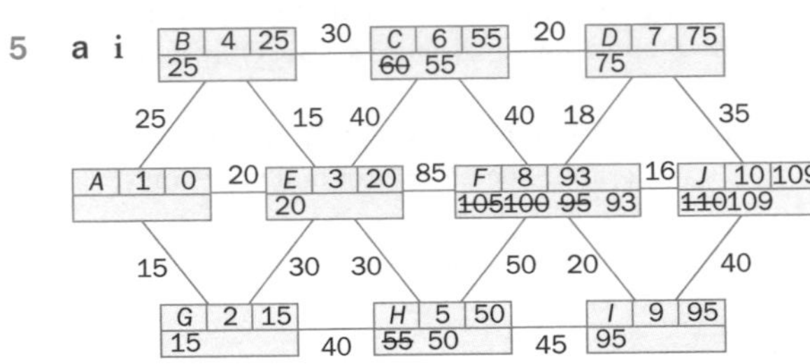

ii *ABCDFJ*, 109 mins

b e.g. *ABCDJ* is 140 mins, while *ABCDFJ* is now 149 mins.

c *AEFJ*, 161 mins (only 2 changes, where all other routes have at least 3).

6 a Kruskal – at each stage choose arc with minimum weight unless it forms a cycle.
Hard to use with table because difficult to spot cycles.
Prim – at each stage choose arc of minimum weight joining a chosen node to one not yet chosen. Cycles impossible, so can be used with table.

b i Starting from *A*: *AC*, *AB*, *BD*, *BE*, *EF*, *EG*

ii *EF*, *AC*, *BD*, *AB*, (not *AD*), *EG*, (not *FG*), *BE*

Chapter 6

Exercise 6.1

1 a

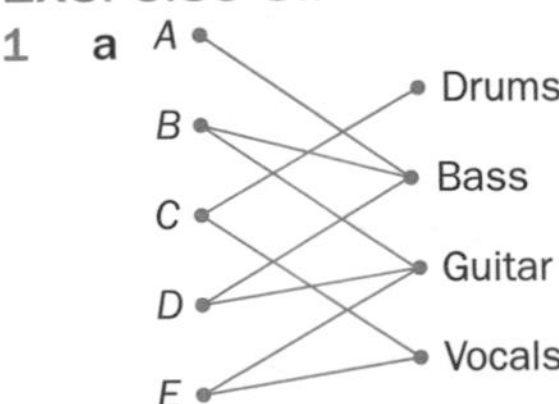

b Ali – bass, Ben or Dee – guitar, Cass – drums, Eve – vocals

2

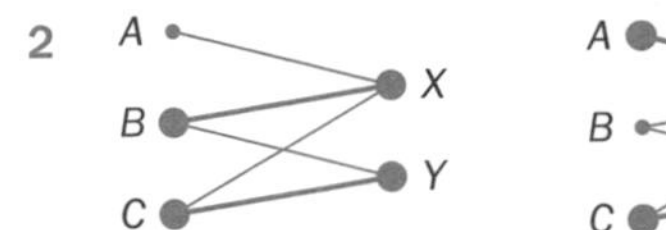

3

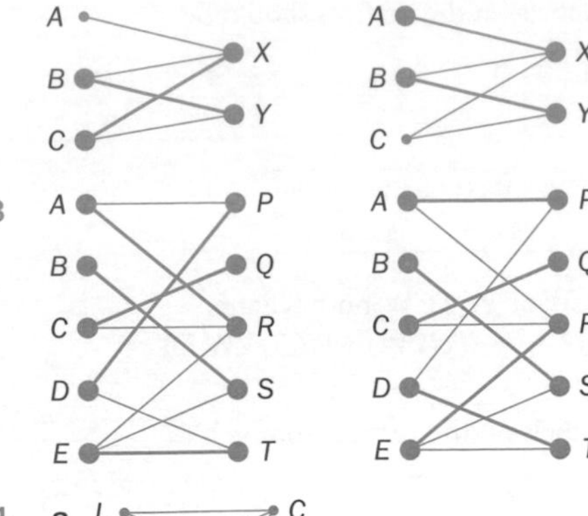

4 a

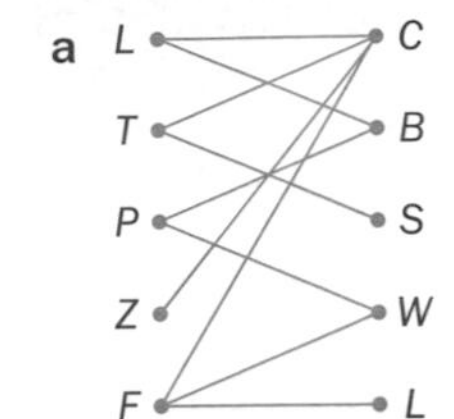

b Matching *FL*, *ZC*, *LB*, *TS*, *PW*.
All choices forced, so this is the only possible matching.

5 a

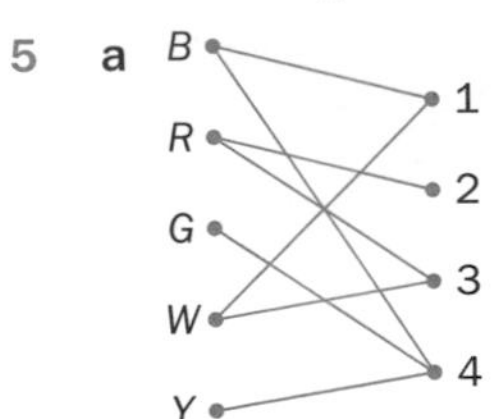

b 1 – *B*, 2 – *R*, 3 – *W*, 4 – *Y*

c Green and yellow

Exercise 6.2

1 a Alternating paths *C*-2+*A*-1+*D*-3+*B*-5 or *C*-1+*D*-3+*B*-5, giving matchings {*A*1, *B*5, *C*2, *D*3, *E*4} or {*A*2, *B*5, *C*1, *D*3, *E*4}

b Maximal

c Alternating path *D*-4+*A*-2. Max. matching {*A*2, *B*1, *C*3, *D*4}

d Alternating paths *C*-1+*B*-3 and *F*-6+*A*-2+*E*-5. Max. matching {*A*2, *B*3, *C*1, *D*4, *E*5, *F*6}

e Alternating path *D*-3+*E*-5. Max. matching {*B*4, *C*1, *D*3, *E*5}

2

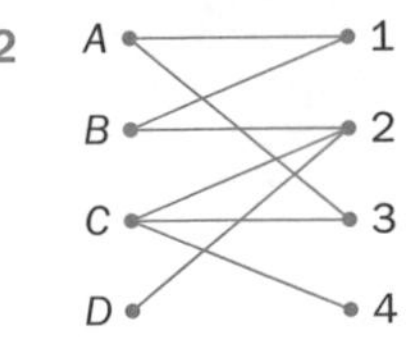

Alternating path *A*-3+*C*-4
Complete matching {*A*3, *B*1, *C*4, *D*2}

3

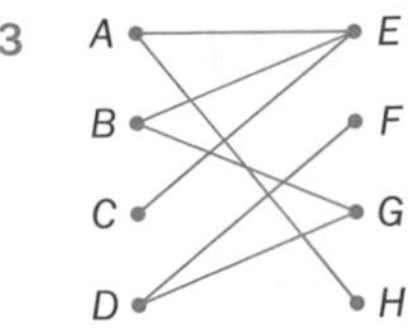

Alternating path *C*-*E*+*A*-*H*.
Complete matching {*AH*, *BG*, *CE*, *DF*}

4 a, b

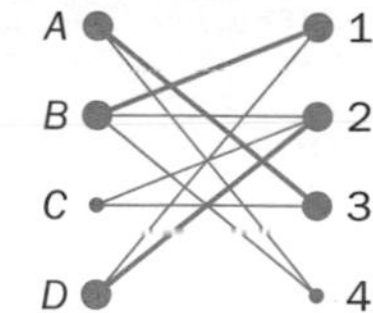

c Path *C*-3+*A*-4, matching {*A*4, *B*1, *C*3, *D*2}
Path *C*-2+*D*-1+*B*-4, matching
{*A*3, *B*4, *C*2, *D*1}

5 a

Possible matchings {*A*3, *B*2, *C*1, *D*4}, {*A*1, *B*2, *C*4, *D*3} or {*A*3, *B*1, *C*4, *D*2}

b Cheapest is {*A*3, *B*2, *C*1, *D*4}, cost £980.

Review 6

1 a

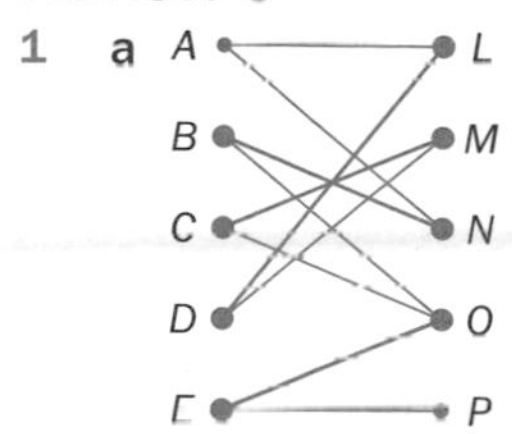

b Initial matching as shown.
Paths *A*-*L*+*D*-*M*+*C*-*O*+*E*-*P* or *A*-*N*+*B*-*O*+*E*-*P*
Matchings {*AL*, *BN*, *CO*, *DM*, *EP*} or {*AN*, *BO*, *CM*, *DL*, *EP*}

2 a

b Paths *K*-*F*+*J*-*E*+*M*-*H* or *K*-*G*+*L*-*F*+*J*-*E*+*M*-*H*
Matchings {*JE*, *KF*, *LG*, *MH*} or {*JE*, *KG*, *LF*, *MH*}

3 a

b Initial matching as shown.
Path *E*-*S*+*C*-*P* gives matching {*AQ*, *CP*, *DT*, *ES*}
No further alternating path, so maximal.

c Another solution is given by the path *E*-*T*+*D*-*R*, with matching {*AQ*, *CS*, *DR*, *ET*}

4 a

b At least two possible pairs of alternating paths:
S-3+*J*-2+*A*-1+*D*-5, *T*-2+*A*-3+*J*-4+*P*-6 or
S-3+*J*-4+*P*-6, *T*-2+*A*-1+*D*-5
All give matching {*A*1, *D*5, *J*4, *P*6, *S*3, *T*2}

5 a

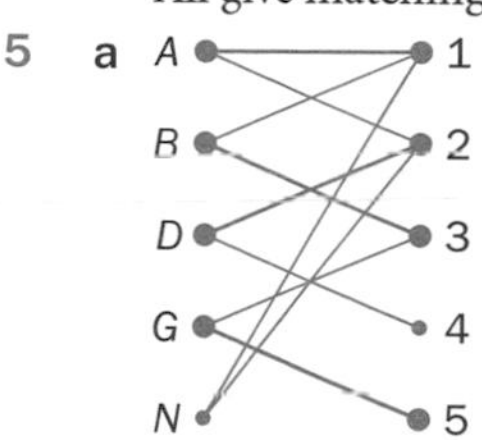

b *N*-1+*A*-2+*D*-4 giving {*A*2, *B*3, *D*4, *G*5, *N*1}

c *N*-2+*D*-4 giving {*A*1, B3, *D*4, *G*5, *N*2}

6 a {1*C*, 2*B*, 3*A*, 4*E*, 5*D*}, {1*C*, 2*A*, 3*D*, 4*B*, 5*E*}

b 2-*D*+5-*E*, 2-*B*+4-*C*+1-*E*

Chapter 7

Check in

1

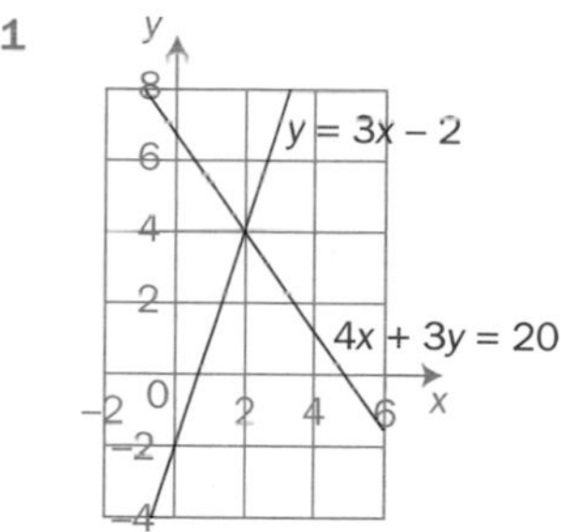

2

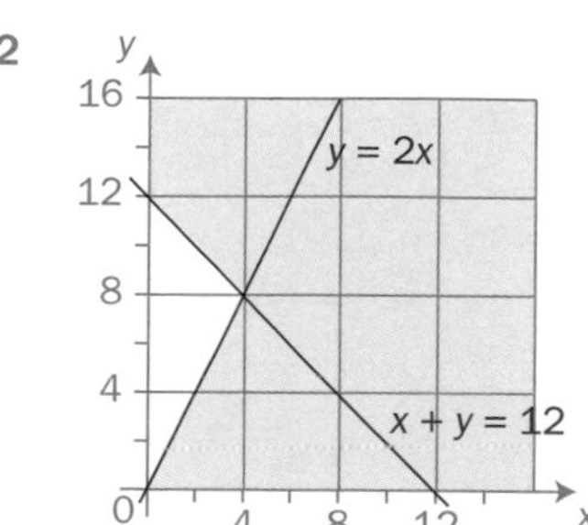

3 $x = 1, y = 2$

Exercise 7.1

1 x litres of Econofruit, y litres of Healthifruit.
Maximise $30x + 40y$ subject to $0.2x + 0.4y \leqslant 20000$, $0.5x + 0.3y \leqslant 30000$, $x \geqslant 0, y \geqslant 0$

2 x cars, y minibuses.
Minimise $20x + 60y$ subject to $5x + 12y \geqslant 80$, $x + y \leqslant 8$, $x \geqslant 0$, $y \geqslant 0$, x and y are integers.

3 x ha of wheat, y ha of potatoes.
Maximise $80x + 100y$ subject to
$30x + 50y \leqslant 2800$, $700x + 400y \leqslant 40000$,
$x + y \leqslant 75$, $x \geq 0$, $y \geqslant 0$

4 x oranges, y apples, z pears.
Minimise $20x + 12y + 15z$ subject to $x \leqslant y$,
$y \geqslant 2z$, $x + y + z \geqslant 30$, $x > 0$, $y > 0$, $z > 0$,
x, y, z are integers.

5 x millilitres of whisky, y millilitres of ginger wine.
Minimise $0.012x + 0.005y$ subject to $x + y \geqslant 100$,
$0.4x + 0.12y \leqslant 30$, $0.2x \geqslant 0.08y$
$x > 0$, $y > 0$.

6 x tins of Vitapuss, y tins of Pamper.
Maximise $51x + 70.75y$ subject to
$0.2x + 0.15y \leqslant 4000$, $0.3x + 0.35y \leqslant 6500$,
$x \geqslant 0$, $y \geqslant 0$, x and y integers.

7 x cases of shampoo, y of cleaner.
Maximise $20x + 14y$ subject to $x + y \leqslant 60$,
$6x + 4y \leqslant 280$, $x \geqslant 0$, $y \geqslant 0$, x and y integers.

8 x bonds, y shares.
Maximise x subject to $x + y \leqslant 20000$,
$0.05x + 0.1y \geqslant 0.08(x + y)$
(which simplifies to $2y \geqslant 3x$), $x \geqslant 0$,
$y \geqslant 0$, x and y integers.

9 x kg Regular mix, y kg Luxury mix
Maximise $1.7x + 2.5y$
subject to $0.7x + 0.5y \leqslant 8000$
$0.3x + 0.5y \leqslant 6000$
$x \geqslant 0$, $y \geqslant 0$

10 Maximise $x + y$ subject to
$6x + 9y \leqslant 45\,000$ (i.e. $2x + 3y \leqslant 15\,000$)
$\frac{x}{4} + \frac{y}{6} \leqslant 1200$ (i.e. $3x + 2y \leqslant 14\,400$)
$x \geqslant 0$, $y \geqslant 0$

11 x rolls of Floral, y rolls of Modern and
z rolls of Swirl per day
Maximise $60x + 40y + 50z$
subject to $2x + y + 2z \leqslant 40$
$3x + 2y + 2z \leqslant 60$
$x + 2y + 2z \leqslant 30$
$x \geqslant 0$, $y \geqslant 0$, $z \geqslant 0$
x, y, and z integers

Exercise 7.2

1 a

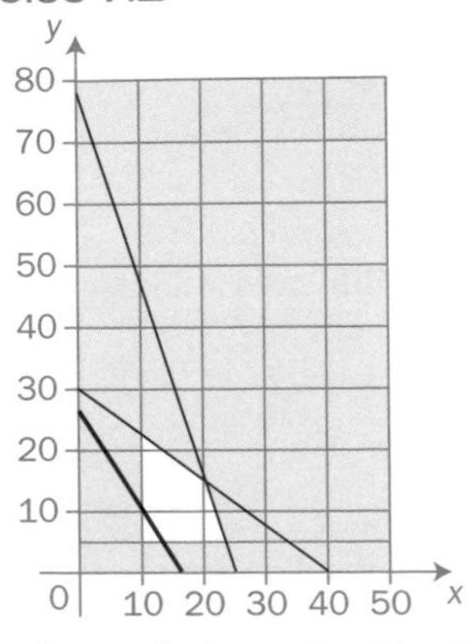

The graph shows the objective line at $P = 50$.
Optimal value $P = 90$, $x = 20$, $y = 15$

b

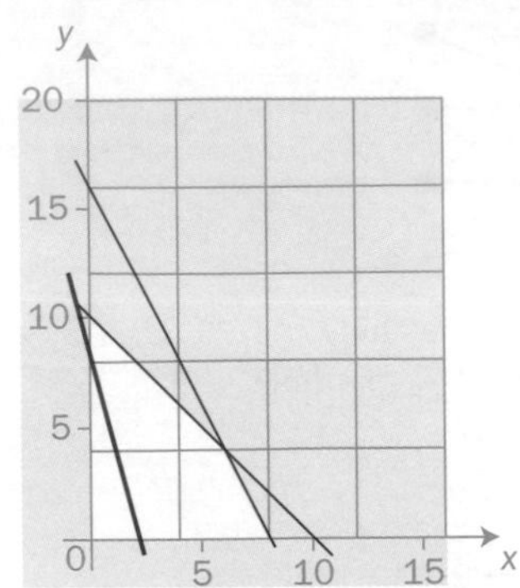

The graph shows the objective line at $P = 8$.
Optimal value $P = 32$, $x = 8$, $y = 0$.

c

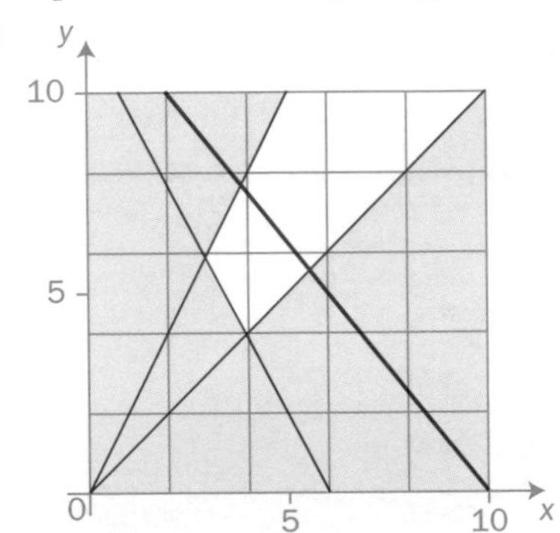

The graph shows the objective line at $C = 50$.
Optimal value $C = 36$ at $x = 4$, $y = 4$.

d

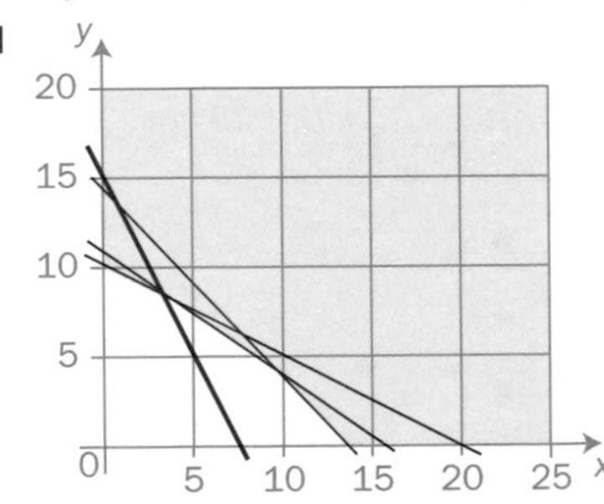

The graph shows the objective line at $R = 15$.
Optimal value $R = 28$, $x = 14$, $y = 0$.

2

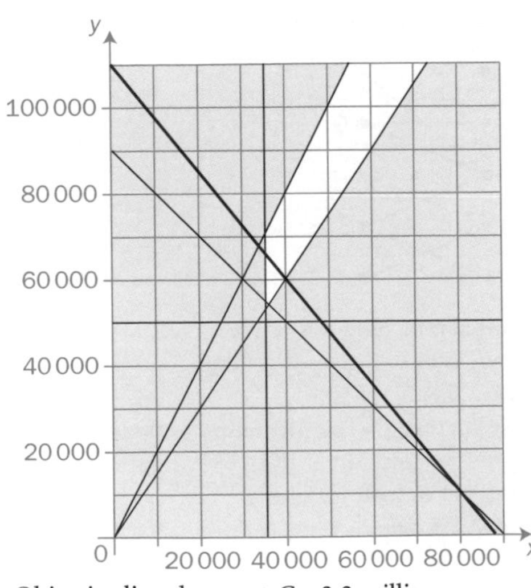

Objective line drawn at $C = 2.2$ million.
Optimal $x = 35\,000$, $y = 55\,000$

3

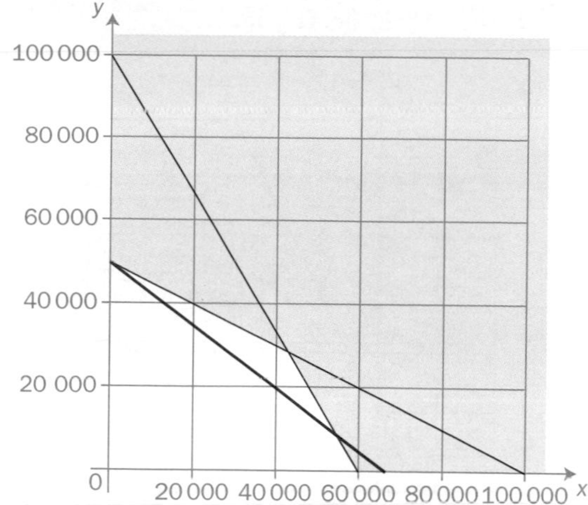

Objective line drawn at $P = 2\,000\,000$.
Optimal value $P = 2\,428\,571\frac{3}{7}$, $x = 42\,857\frac{1}{7}$, $y = 28\,571\frac{3}{7}$

4

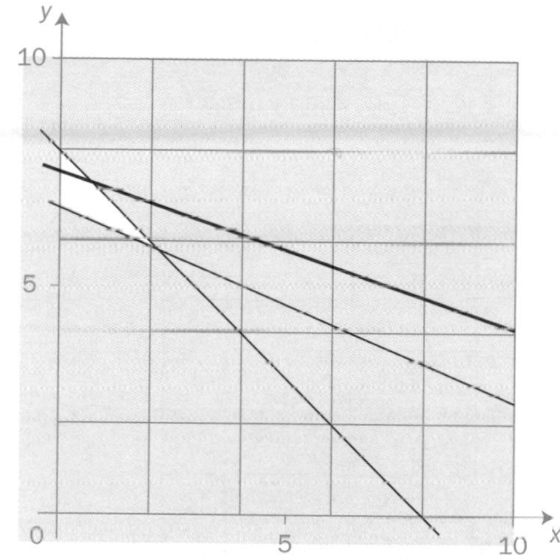

Objective line drawn at $C = 450$.
Optimal value $C = 400$, $x = 2$, $y = 6$.

5

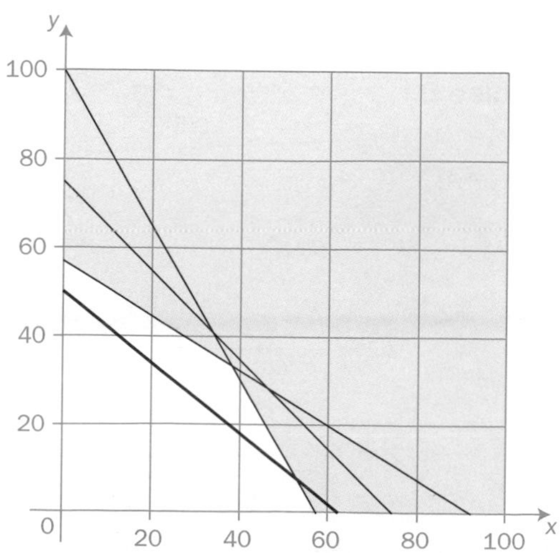

Objective line drawn at $P = 5000$. Optimal value $P = 6340$, $x = 38$, $y = 33$.(Note that the constraint $x + y \leqslant 75$ was redundant.)

6

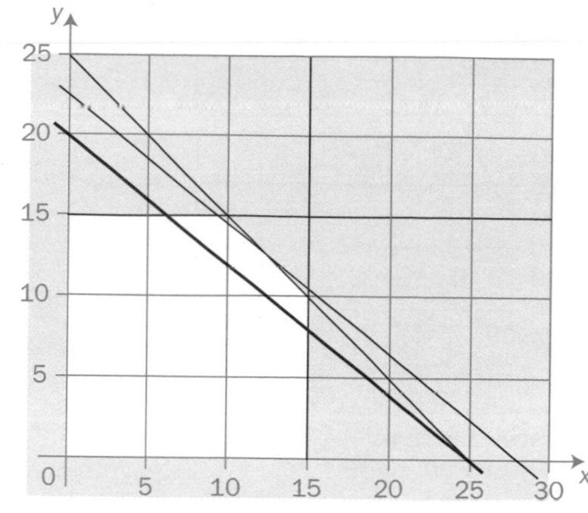

x houses, y bungalows.
Maximise $P = 20\,000x + 25\,000y$ subject to
$x + y \leqslant 25$, $210x + 270y \leqslant 6000$,
$x \leqslant 15$, $y \leqslant 15$, $x \geqslant 0$, $y \geqslant 0$,
x, y are integers.
Objective line drawn at $P = 500\,000$.
Optimal value $P = 560\,000$, $x = 13$, $y = 12$.

7 **a** Minimise $C = 20x + 15y + 12z$ subject to $x + y + z = 50$, $2x + 2y \geqslant z$, $x \geqslant y$, $x \geqslant 10$, $y \geqslant 5$, $x, y, z \geqslant 0$, x, y, z are integers.

c

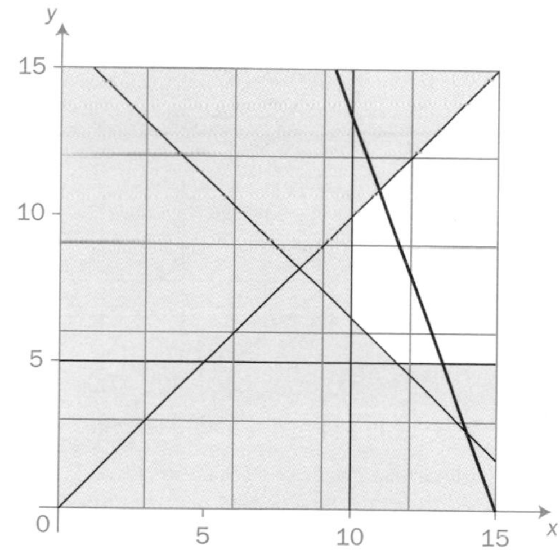

Objective function drawn at $C = 720$.
Optimal value $C = 701$, $x = 10$, $y = 7$, $z = 33$.

8 x of Chew A, y of B, z of C.
Minimise $1.8x + 1.6y + 1.5z$ subject to $x + y + z = 60$, $x \leqslant 30$, $y \leqslant 30$,
$x \geqslant 0$, $y \geqslant 0$
$z \leqslant 30$, $8x + 6y + 5z \leqslant 500$,
$2x + 4y + 5z \leqslant 250$.
This reduces to:
Minimise $0.3x + 0.1y + 90$, subject to
$x \leqslant 30$, $y \leqslant 30$, $x + y \geqslant 30$, $3x + y \leqslant 200$,
$3x + y \geqslant 50$, $x \geqslant, y \geqslant 0$ (x, y, z are integers)

D1

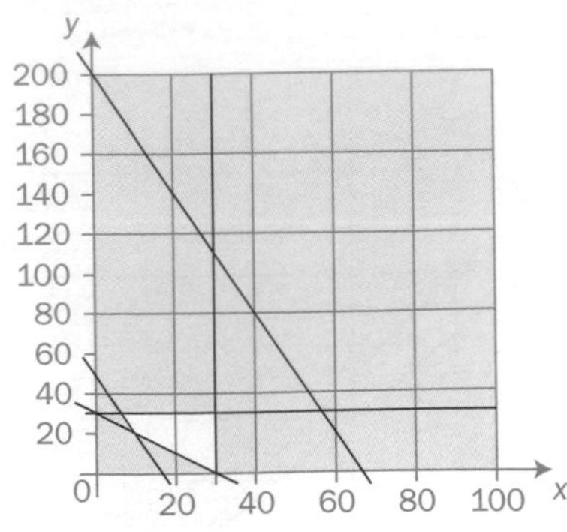

Possible solutions

$x = 10, y = 20, z = 30$

$x = 9, y = 23, z = 28$

$x = 8, y = 26, z = 26$

$x = 7, y = 29, z = 24$

All give a cost of 95p per packet.

Review 7

1 **a** x Value, y Luxury.
Maximise $10x + 12y$ subject to $2x + 3y \leqslant 2400$, $4x + 3y \leqslant 4050$, $x \geqslant 0$, $y \geqslant 0$.

b

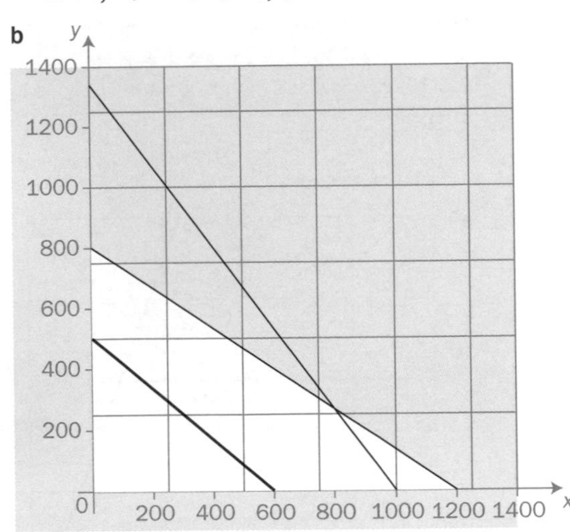

c Max profit £112.50, $x = 825$, $y = 250$

2 **a** Maximise $P = 2.5x + 3y$, subject to $x + 2y \leqslant 30$, $2x + y \leqslant 35$, $x + y \leqslant 20$, x and y integers, $x \geqslant 0$, $y \geqslant 0$.

b

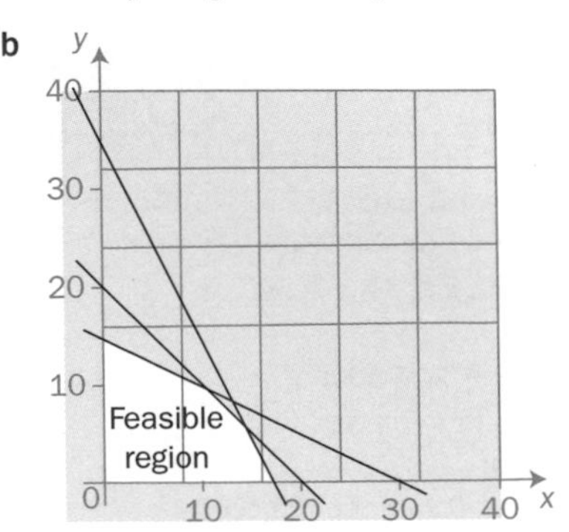

c Max $P = £55$, with $x = 10$, $y = 10$.

d Dept B has 5 hours spare.

3 **a**

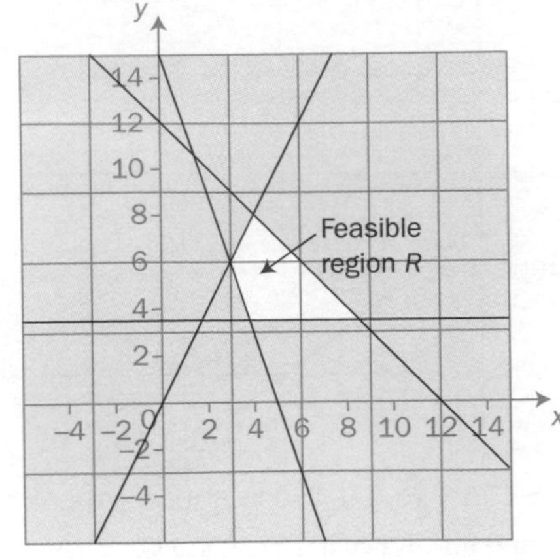

b $x = 3\frac{5}{6}, y = 3\frac{1}{2}, C = 25\frac{1}{6}$

c $x = 8, y = 4, P = 32$

4 **a** Maximise $30x + 40y$ subject to $x + y \leqslant 500$, $x + y \geqslant 200$, $4x \geqslant y$, $3x \leqslant 2y$, x and y integers, $x \geqslant 0$, $y \geqslant 0$.

b

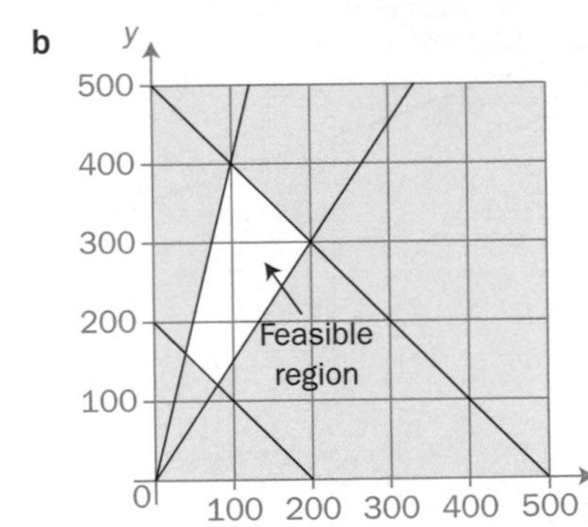

c $x = 100$, $y = 400$, profit £190.

Chapter 8

Exercise 8.1

1 **a**

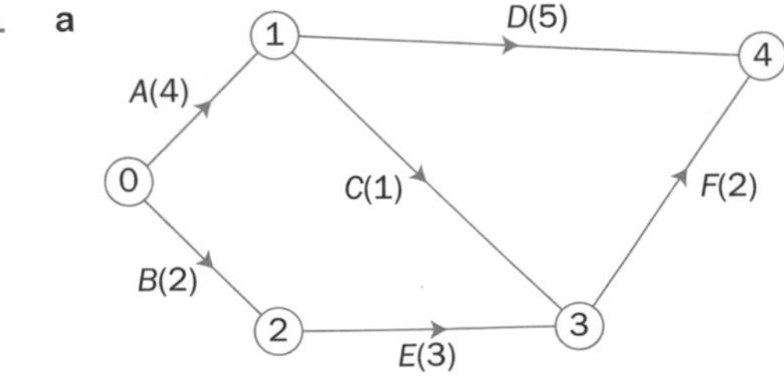

b

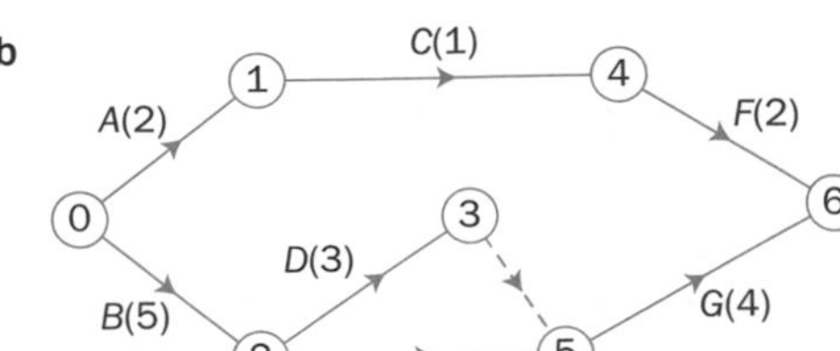

D1

c

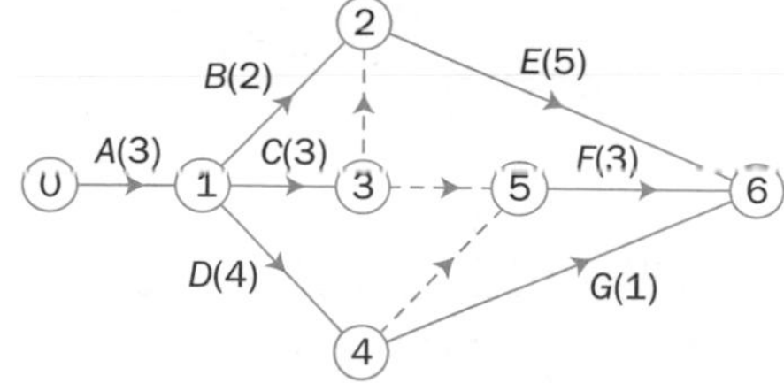

d

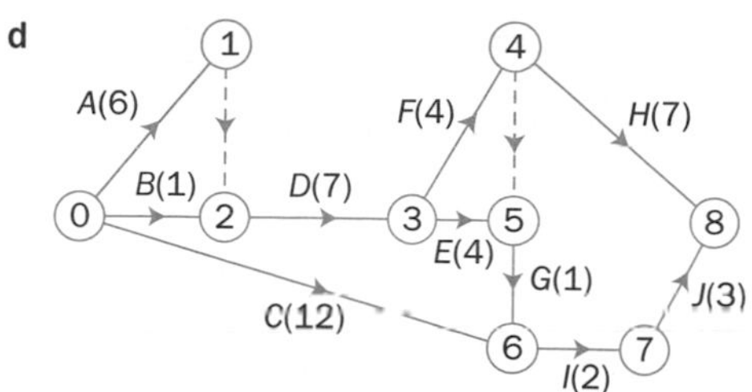

e

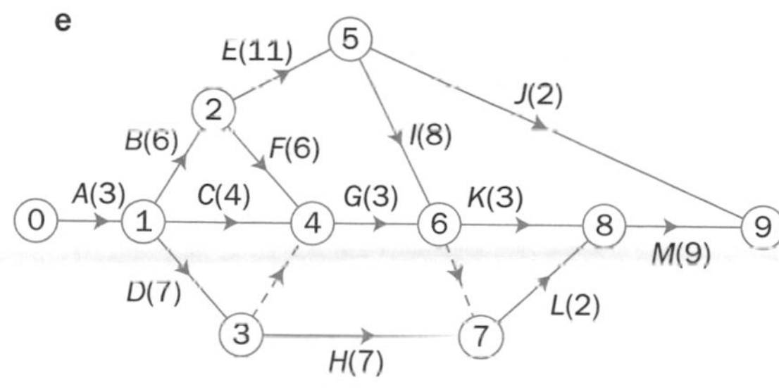

2

Activity	Depends on
A	–
B	–
C	–
D	*A*
E	*A*
F	*D*, *E*
G	*B*, *E*
H	*C*, *D*

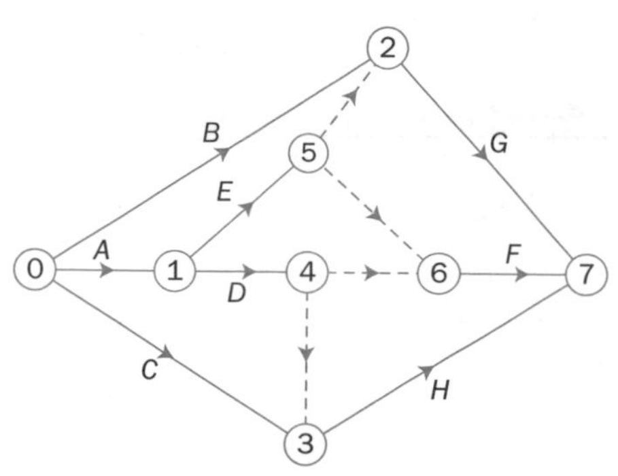

3

Activity	Depends on
A(0,1)	–
B(1,2)	*A*
C(1,3)	*A*
D(2,6)	*B*
E(2,4)	*B*
F(3,4)	*C*
G(3,5)	*C*
H(4,7)	*E*, *F*
I(5,8)	*G*
J(6,10)	*D*
K(7,9)	*D*, *G*, *H*
L(8,9)	*E*, *F*, *I*
M(9,10)	*K*, *L*
N(8,10)	*E*, *F*, *I*

Exercise 8.2

1 a

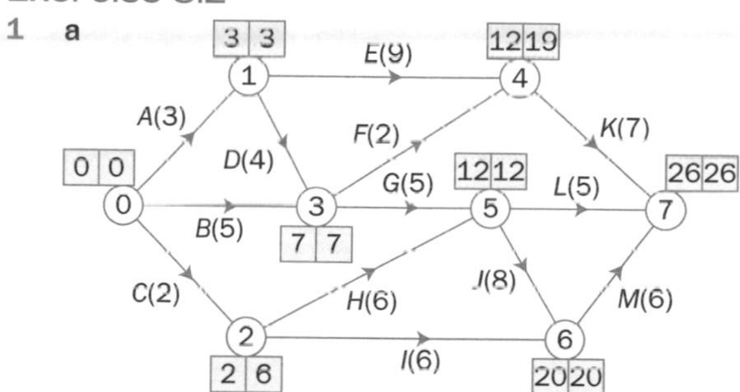

Project duration 26 days. Critical activities *A*, *D*, *G*, *J*, *M*.

b

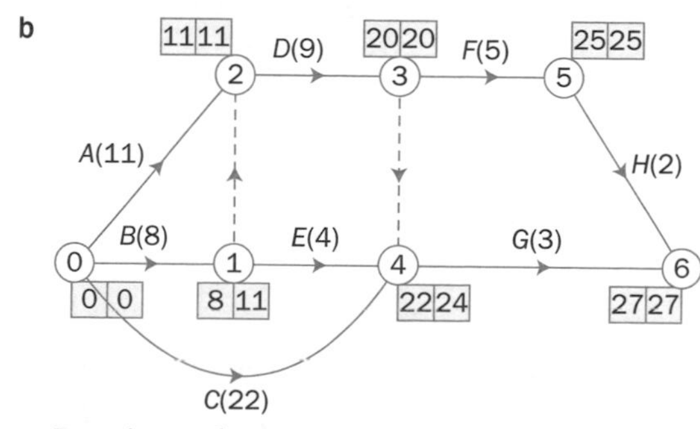

Duration 27 days. Critical activities *A*, *D*, *F*, *H*.

c

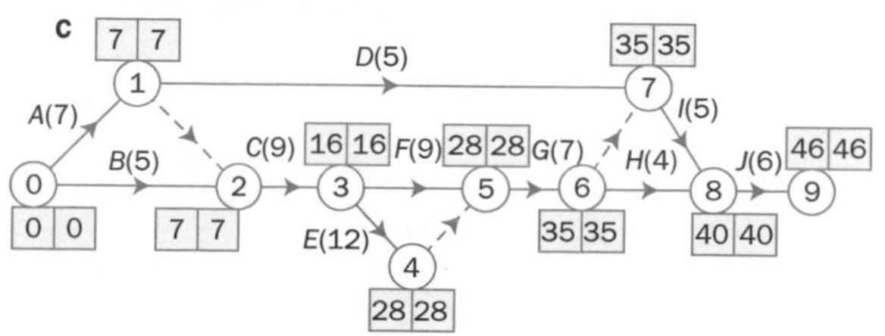

Duration 46 days. Critical activities *A*, *C*, *E*, *G*, *I*, *J*.

D1

d

Duration 24 days. Critical activities *B*, *D*, *E*, *F*, *G*, *H*.

2 **a**

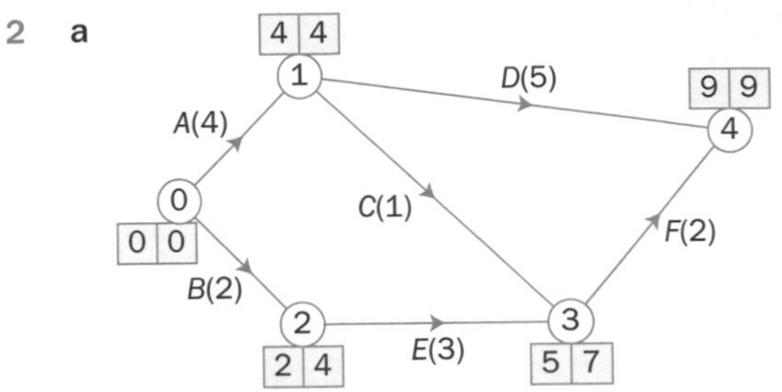

Critical activities *A*, *D*.

b

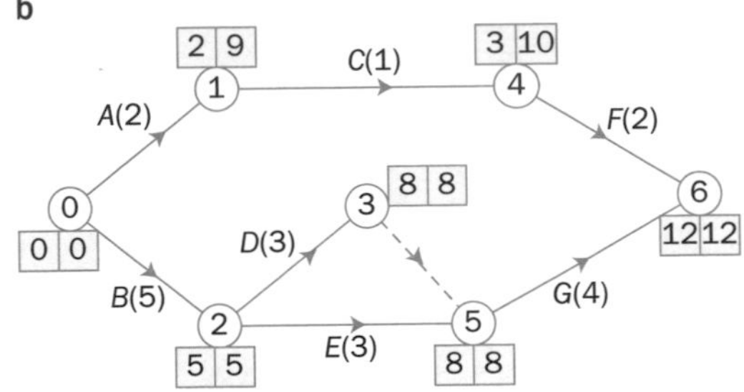

Critical activities *B*, *D*, *E*, *G*.

c

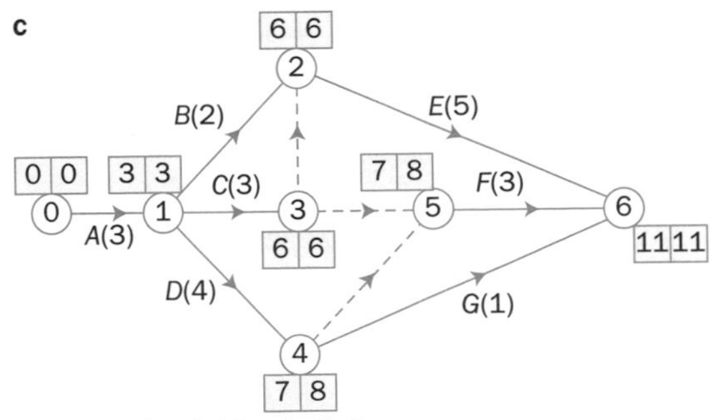

Critical activities *A*, *C*, *E*.

d

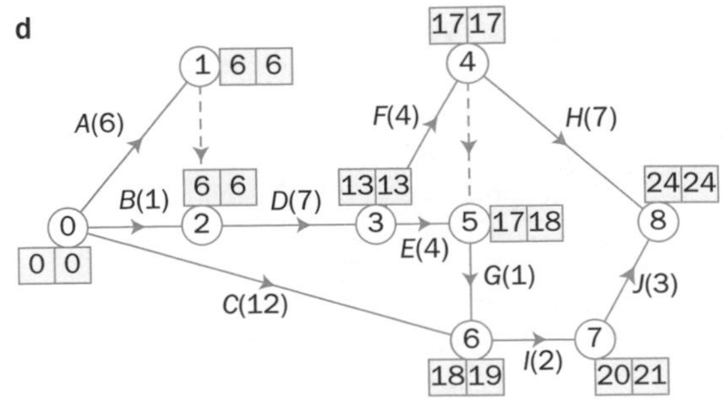

Critical activities *A*, *D*, *F*, *H*.

e

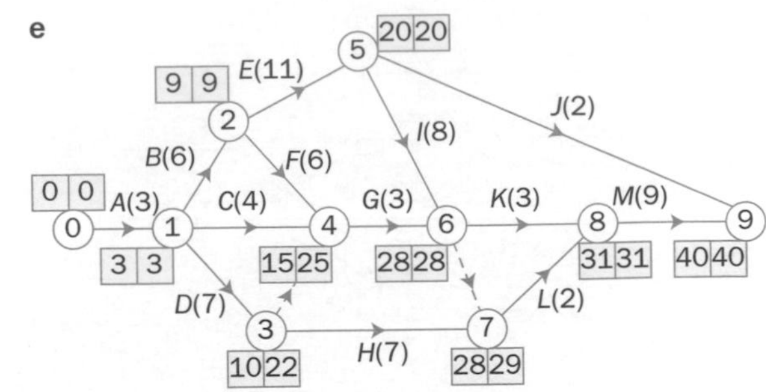

Critical activities *A*, *B*, *E*, *I*, *K*, *M*

3

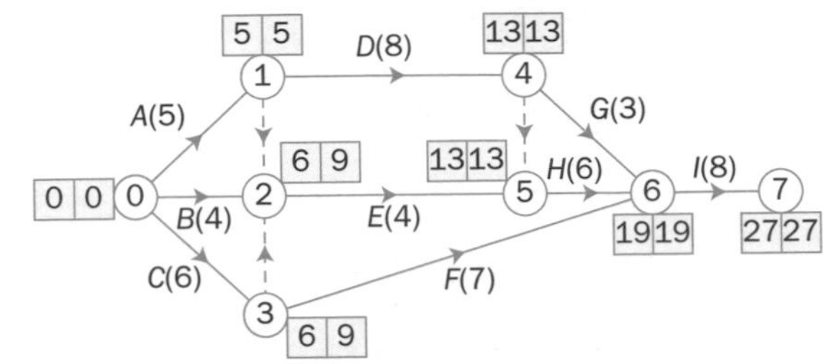

a Duration 27 hours

b *A*, *D*, *H*, *I*

c *B*5, *C*3, *E*3, *F*6, *G*3

Exercise 8.3

1

Activity	Duration (hours)	Start		Finish		Float
		Earliest	Latest	Earliest	Latest	
A	6	0	0	6	6	0
B	8	0	7	8	15	7
C	14	6	8	20	22	2
D	12	6	6	18	18	0
E	3	8	15	11	18	7
F	4	18	18	22	22	0
G	3	22	22	25	25	0

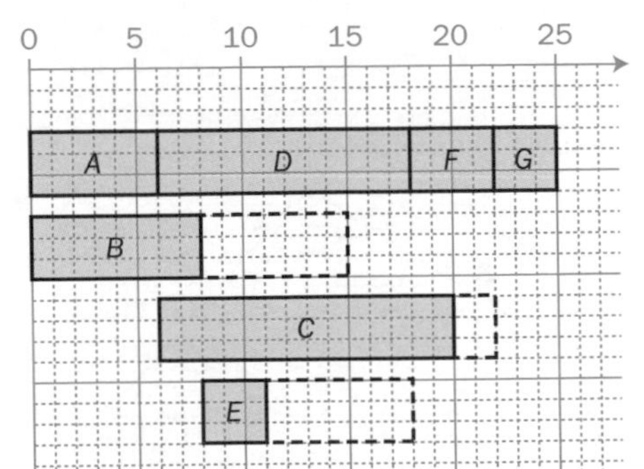

2

3

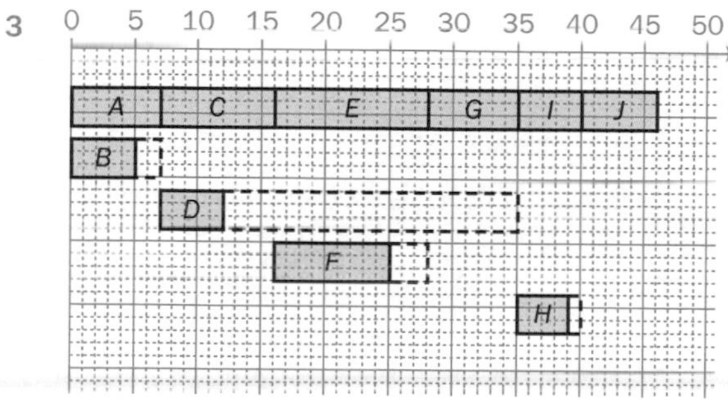

4

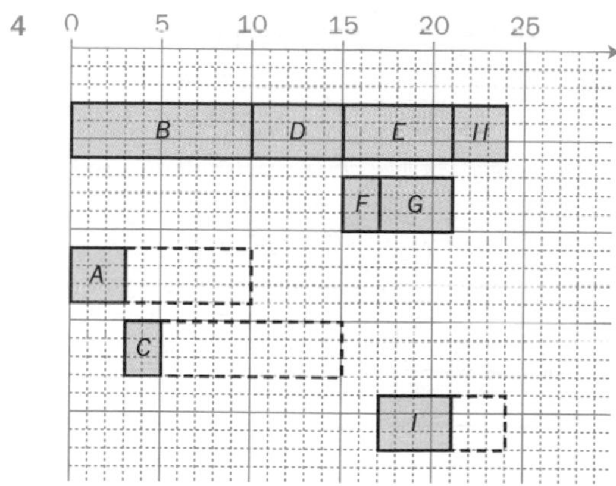

5 a

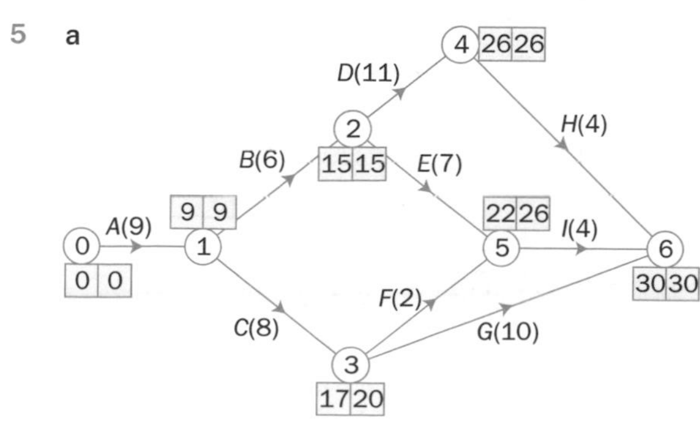

b

Activity	Duration (hours)	Start		Finish		Float
		Earliest	Latest	Earliest	Latest	
A	9	0	0	9	9	0
B	6	9	9	15	15	0
C	8	9	12	17	20	3
D	11	15	15	26	26	0
E	7	15	19	22	26	4
F	2	17	24	19	26	7
G	10	17	20	27	30	3
H	4	26	26	30	30	0
I	4	22	26	26	30	4

c

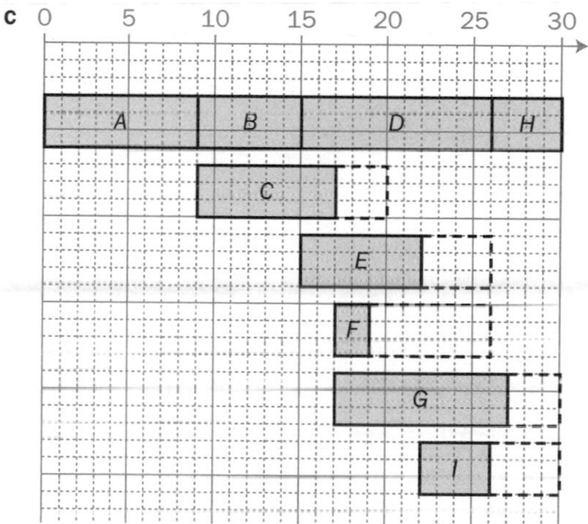

d Activity *B* by 3 hours = £150 (above this *C* becomes critical).

Exercise 8.4

1 a 3 workers because *C*, *D* and *E* have to be done in parallel.

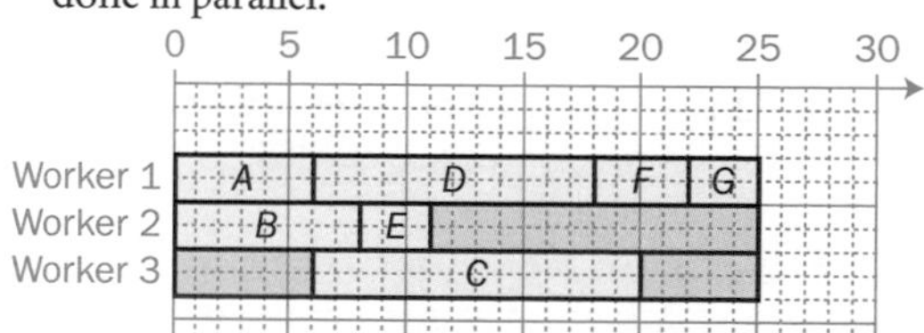

b Duration 28 hours.

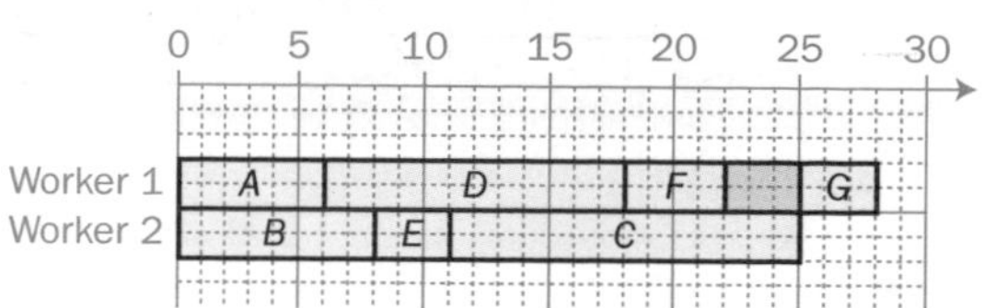

2 a

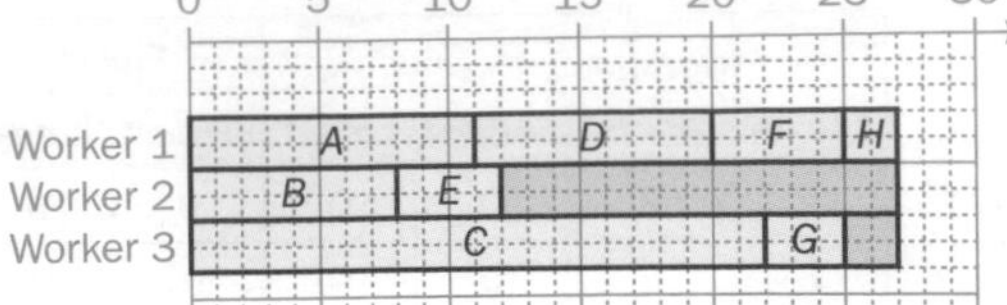

In total, workers are idle for 17 days.

b Worker 2 is idle after day 12. Either Worker 2 leaves (no effect) or Worker 2 takes over from the leaver.

3 a

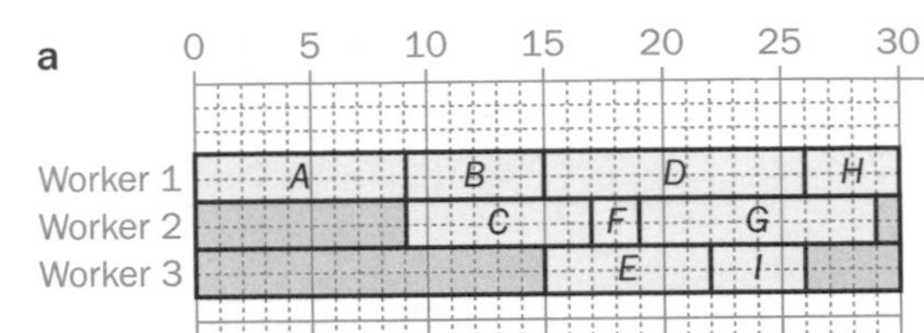

b *E* and *I* could be 19-26 and 26-30, so one extra worker after 19 days.

4 a

Activity	Duration (weeks)	Start		Finish		Float
		Earliest	Latest	Earliest	Latest	
A	7	0	5	7	12	5
B	7	0	5	7	12	5
C	2	0	0	2	2	0
D	3	2	2	5	5	0
E	2	2	3	4	5	1
F	7	5	5	12	12	0
G	2	4	20	6	22	16
H	8	6	22	14	30	16
I	8	5	22	13	30	17
J	18	12	12	30	30	0

30 weeks is minimum completion time

b Sum = 64 weeks. This is more than twice the duration, so 3 workers needed.

c e.g.

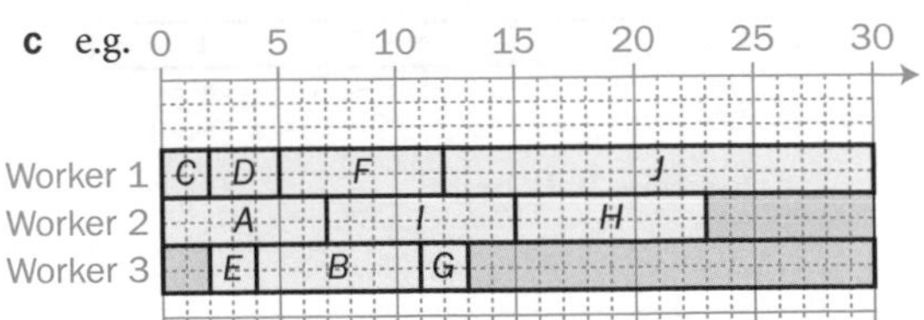

d 32 weeks.

5

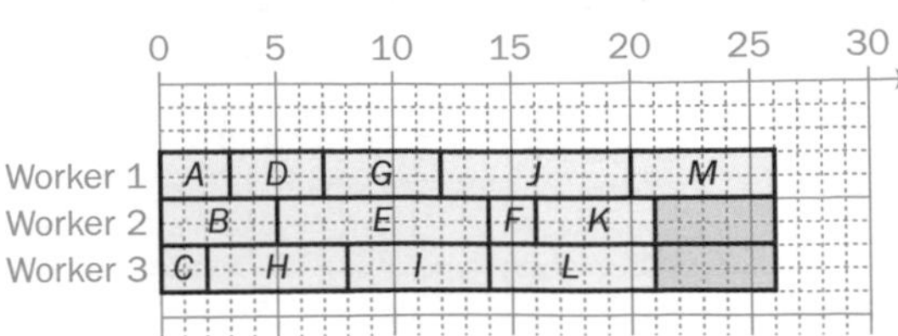

6 a Duration 32 weeks.

b

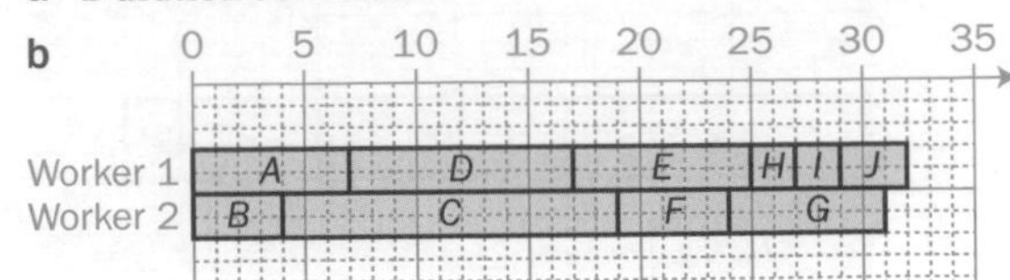

c i Project time reduced to 31 weeks.

ii Project could still be completed with 2 workers.

Review 8

1 a

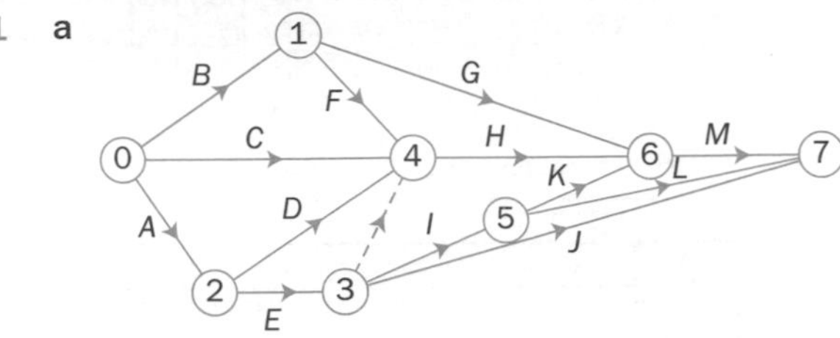

b Because *H* depends on *E* and others, while *I*, *J* depend only on *E*.

2 a

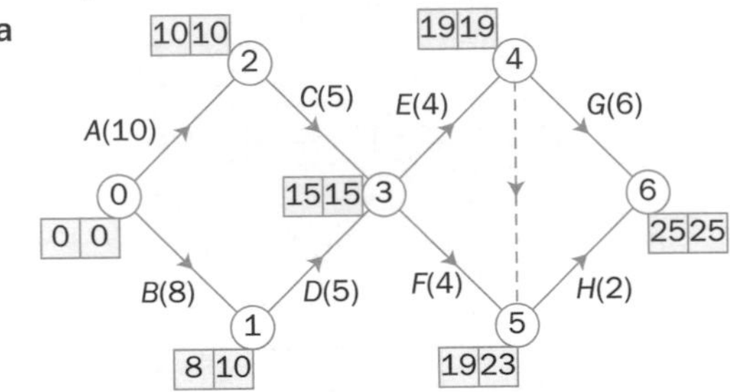

Dummy needed because *G* depends on *E* only, but *H* depends on *E* and *F*.

b i 25 days **ii** *A*, *C*, *E*, *G*

c

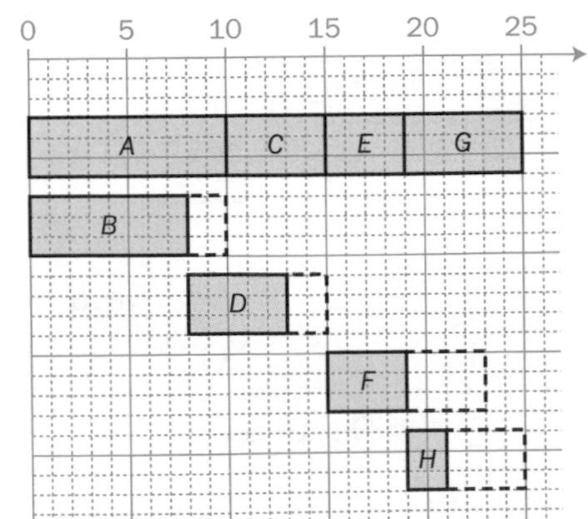

3 a

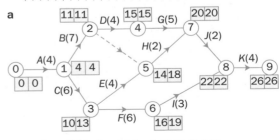

26 days. Critical activities *A*, *B*, *D*, *G*, *J*, *K*

b 4 days

c Can shorten all three without others becoming critical, so spend £300.

4 a

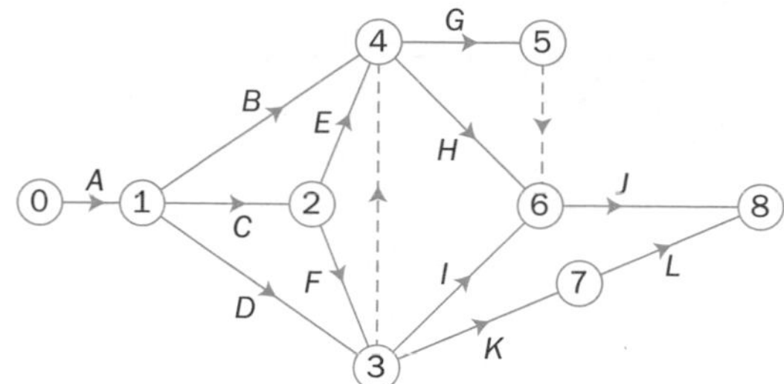

b $x \geqslant 2$

5 a

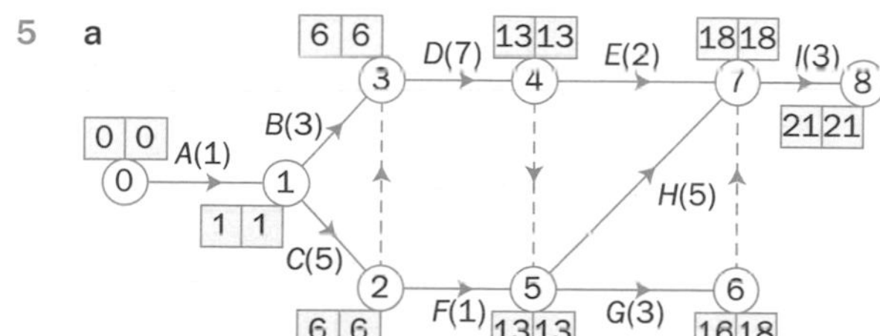

b Time 21 days, critical activities *A, C, D, H, I.*

c *B*2, *E*3, *F*6, *G*2

d

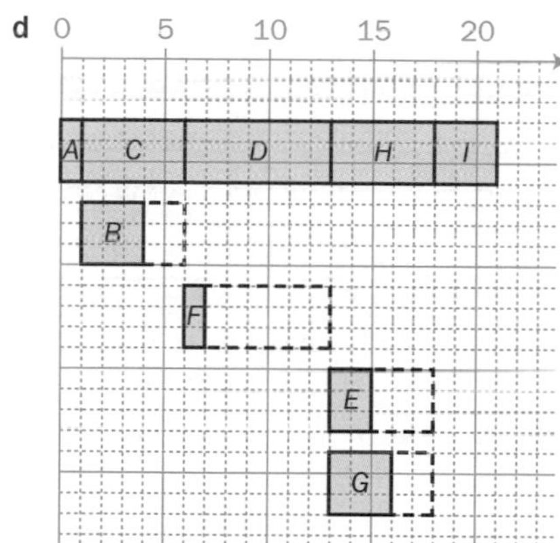

e

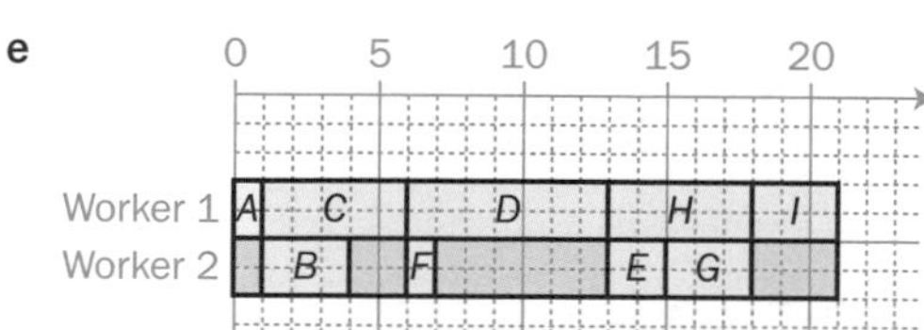

6 a

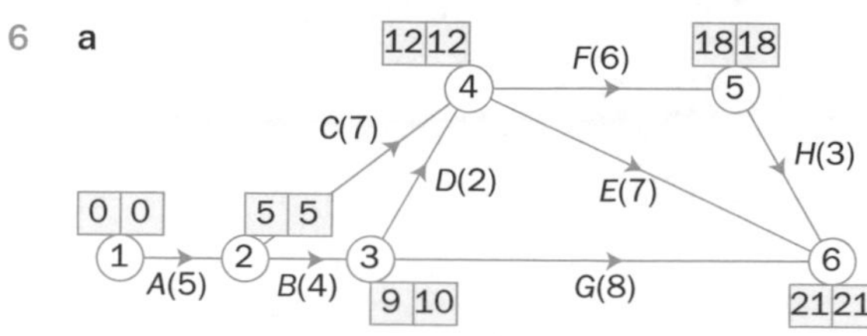

b Critical activities *A, C, F, H.* Duration 21 days.

c *B*1, *D*1, *E*2, *G*4

d

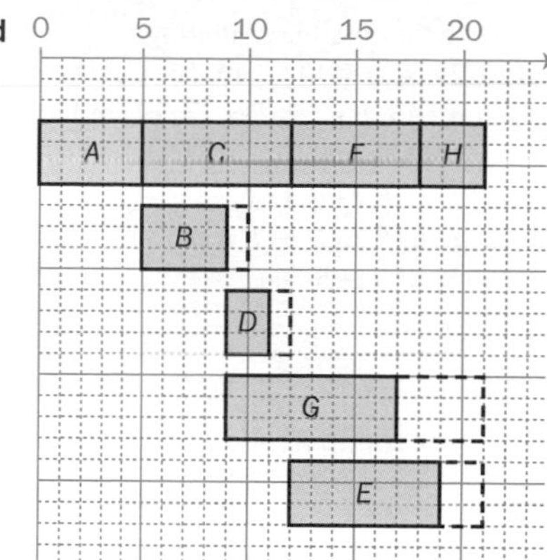

e

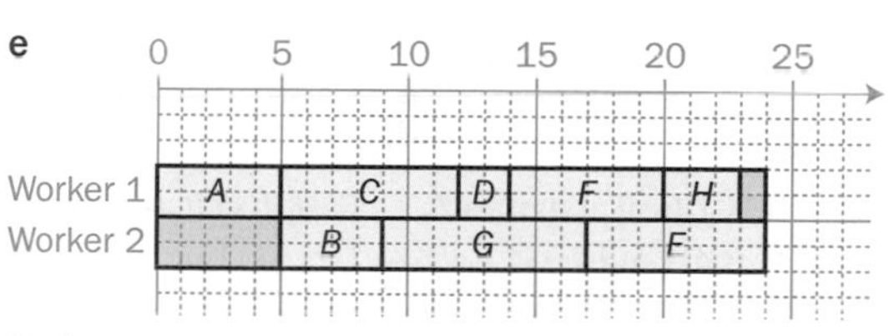

24 days

7 a $x = 0, y = 7, z = 9$

b Length 22 hours, critical activities *B, D, E, L.*

c i 5 hours **ii** 8 hours

d

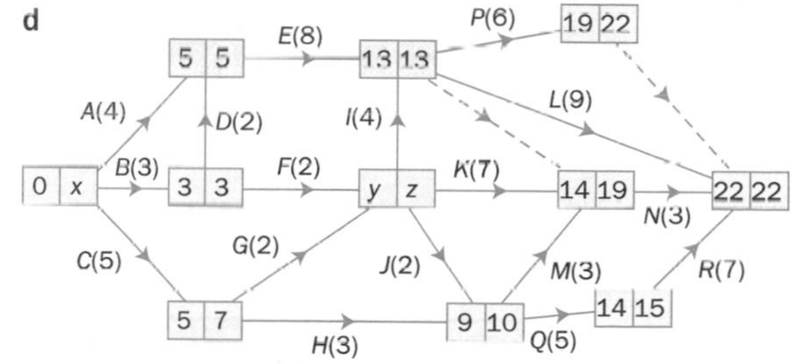

e 22 hours

Revision 2

1 a

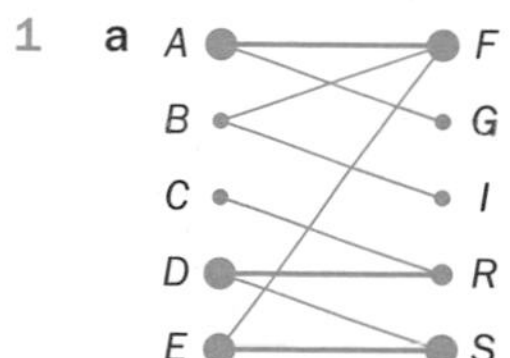

b *B-F+A-G, C-R+D-S+E-F+B-I* or *C-R+D-S+E-F+A-G, B-I*

Both give matching {*AG, BI, CR, DS, EF*}

2 a

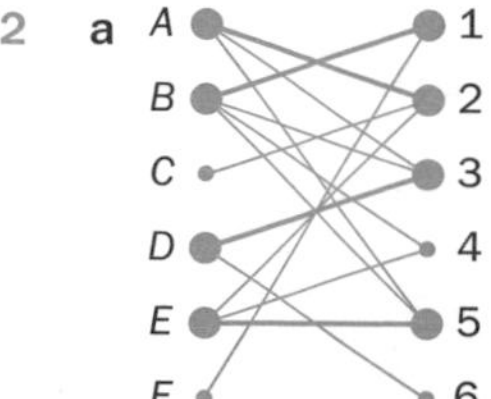

b There are three possible matchings: {$A3$, $B4$, $C2$, $D6$, $E5$, $F1$}, {$A3$, $B5$, $C2$, $D6$, $E4$, $F1$} and {$A5$, $B3$, $C2$, $D6$, $E4$, $F1$}and at least three pairs of alternating paths leading to these:
C-2+A-3+D-6, F-1+B-4 or
C-2+A-3+D-6, F-1+B-5+E-4 or
C-2+A-5+E-4, F-1+B-3+D-6

3 **a**

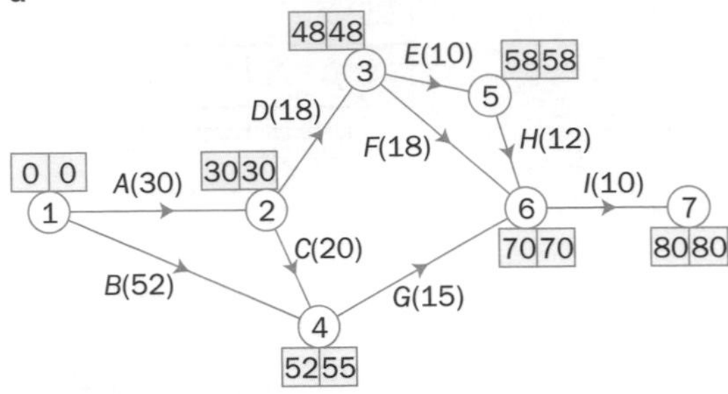

b Critical activities A, D, E, H, I. Duration 80 days.

c $B3$, $C5$, $F4$, $G3$

d

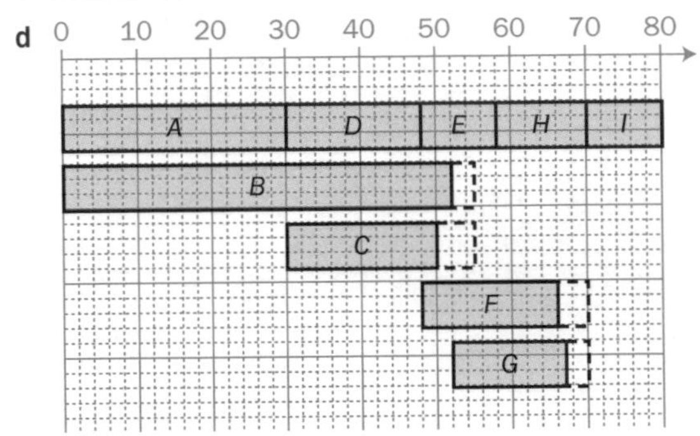

e

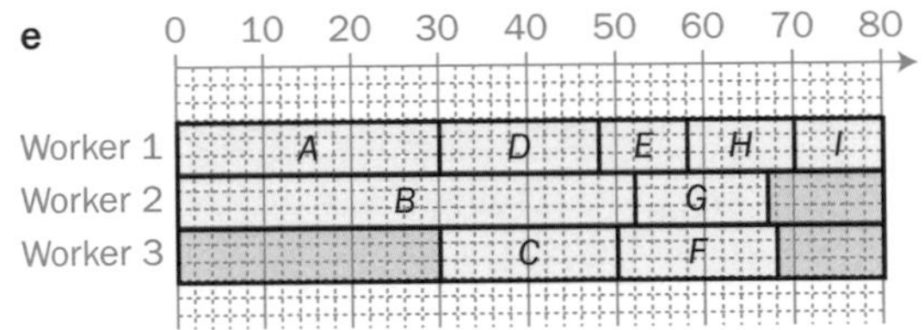

f Duration 85 days, critical activities A, C, G, I.

4 **a, b**

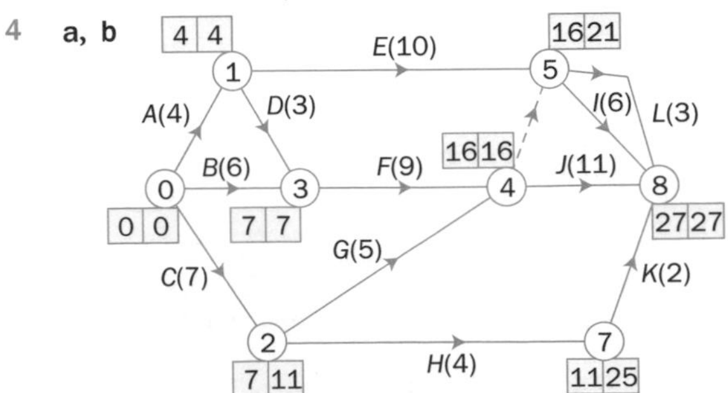

c Critical activities A, D, F, J. Duration 27 days.

d

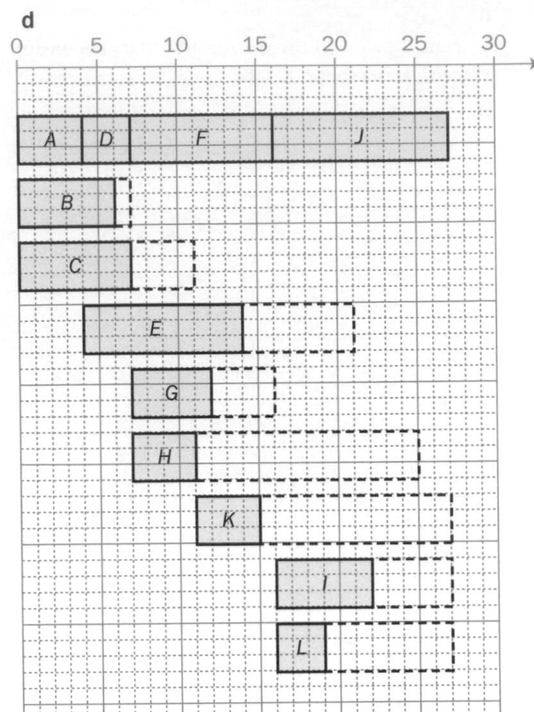

e Minimum is 3 workers.

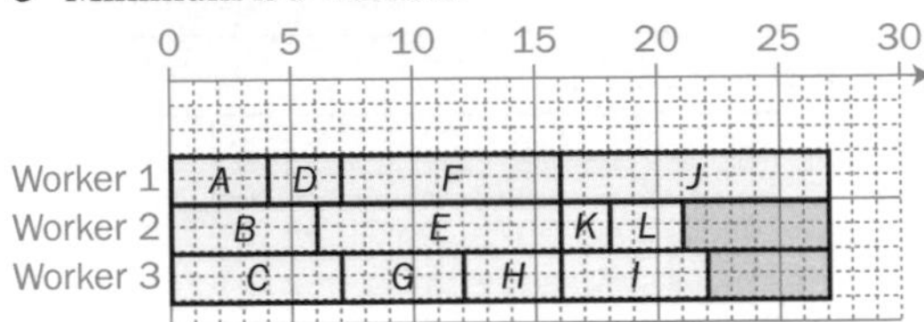

5 **a** Maximise $15x + 15y$ subject to $3x + 10y \leqslant 3000$, $3x + 2y \leqslant 1200$, $x \geqslant 2y$, $x \geqslant 0$, $y \geqslant 0$, x and y are integers.

b

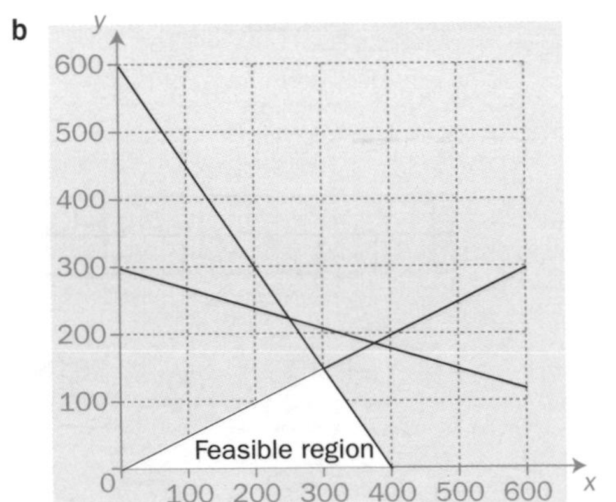

c 300 Infant, 150 Junior, £67.50 profit.

d The sweets constraint does not affect the feasible region. Increase stickers.

e There is not a restricted supply of these

6 a Minimise $T = 2x + 3y$ subject to $5x + y \geqslant 10$, $x + y \geqslant 6$, $x + 4y \geqslant 12$, x and y integers, $x \geqslant 0, y \geqslant 0$

b

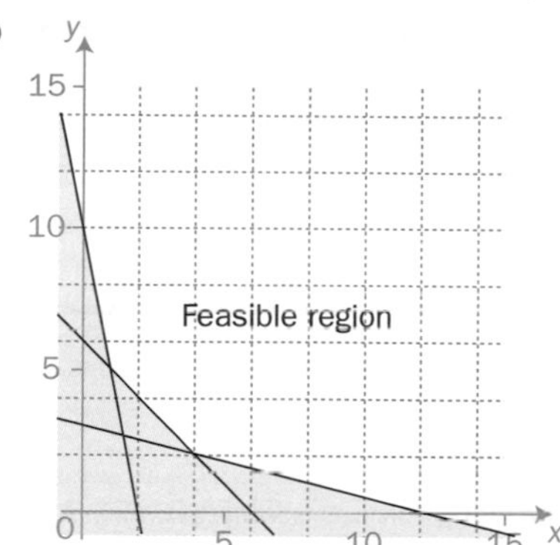

c $x = 4, y = 2$, cost = £14.

7 a

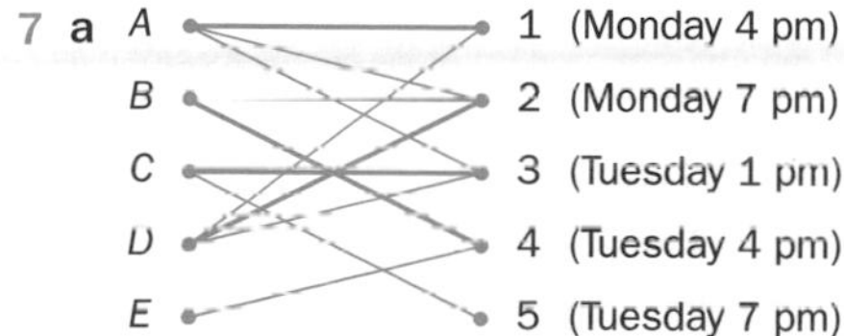

b *E*-4+*B*-2+*D*-3+*C*-5 gives matching {*A*1, *B*2, *C*5, *D*3, *F*4}
or *E*-4+*B*-2+*D*-1+*A*-3+*C*-5 gives {*A*3, *B*2, *C*5, *D*1, *E*4}

c Because *D*2 forces *B*4, leaving *E* out.

8 a

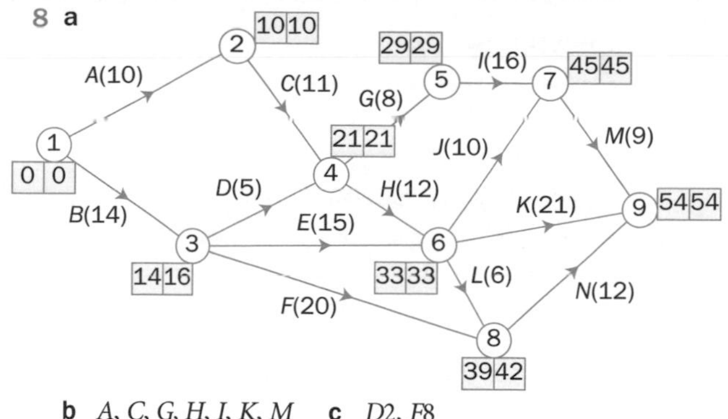

b *A, C, G, H, I, K, M* **c** *D*2, *F*8

d

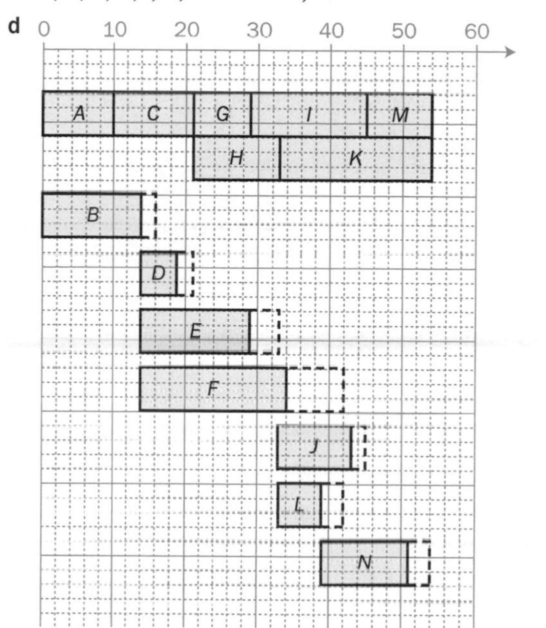

e *C* on day 15, *E, F, G* and *H* on day 25.

Glossary

Keywords are given in bold type.
The black text in this glossary is taken from the Edexcel glossary.
Additional explanation is given in blue type.

Algorithms In a list containing N items the 'middle' item has position $\left[\frac{1}{2}(N+1)\right]$ if N is odd, $\left[\frac{1}{2}(N+2)\right]$ if N is even.

e.g. If $N = 9$, the middle item is the 5th item and if $N = 6$ it is the 4th item.

The notation $\left[\frac{1}{2}(n_1 + n_2)\right]$ means 'the smallest integer greater than or equal to $\frac{1}{2}(n_1 + n_2)$', so in fact the above could just say $\left[\frac{1}{2}(N+1)\right]$ for all N.

In the binary search and quick sort algorithms, the middle position of a list starting at position n_1 and ending at position n_2 is $\left[\frac{1}{2}(n_1 + n_2)\right]$.

A **greedy** algorithm is one which chooses the most advantageous option available at each stage, without looking ahead. Examples of greedy algorithms are the first-fit and first-fit decreasing algorithms for bin packing, and Kruskal's and Prim's algorithms for finding a minimum spanning tree.

Graphs A **graph** G consists of points (**vertices** or **nodes**) which are connected by lines (**edges** or **arcs**).

A graph can be defined by listing its **vertex set** and **edge set**, or by means of an **adjacency matrix** showing the number of connections between pairs of vertices.

A **subgraph** of G is a graph, each of whose vertices belongs to G and each of whose edges belongs to G.

If a graph has a number associated with each edge (usually called its **weight**) then the graph is called a **weighted graph** or **network**.

A network can be defined by means of a **distance matrix** showing the weights of the connections between pairs of vertices.

The **degree** or **valency** (or **order**) of a vertex is the number of edges incident to it. A vertex is **odd (even)** if it has **odd (even)** degree. Every graph has an even number of odd vertices (this is the **hand-shaking lemma**). The sum of the degrees is twice the number of edges.

A **path** is a finite sequence of edges, such that the end vertex of one edge in the sequence is the start vertex of the next, and in which no vertex appears more then once.

If vertices appear more than once the sequence is usually called a **trail**.

Dijkstra's algorithm is used to find the shortest path (path with least total weight) between two chosen vertices.

A **cycle (circuit)** is a closed path, i.e. the end vertex of the last edge is the start vertex of the first edge.

Two vertices are **connected** if there is a path between them. A graph is **connected** if all its vertices are connected (that is, if every possible pair of vertices is connected).

A **simple** graph has no multiple connections (vertices connected by more than one edge) or loops (edges connecting a vertex to itself).

If the edges of a graph have a direction associated with them they are known as **directed edges** and the graph is known as a **digraph** (or **directed graph**).

A **tree** is a connected graph with no cycles.

A **spanning tree** of a graph G is a subgraph which includes all the vertices of G and is also a tree. The spanning tree of a graph with n vertices contains $(n - 1)$ edges.

A **minimum spanning tree** (MST) is a spanning tree such that the total length (total weight) of its arcs is as small as possible. (MST is sometimes called a **minimum connector**.)

A minimum spanning tree can be found using **Kruskal's** or **Prim's** algorithm.

A graph in which each of the n vertices is connected (directly) to every other vertex is called a **complete graph**. The complete graph with n vertices is called K_n. It has $\frac{1}{2}n(n-1)$ edges.

A graph is **traversable** (**Eulerian**) if there is a trail (called an **Eulerian trail**) which starts and ends at the same vertex and which includes every edge once and once only. If there is such a trail but with different start and end points, the graph is **semi-traversable** (**semi-Eulerian**).

The **route inspection problem** (**Chinese postman problem**) is to find the shortest (least weight) closed trail which includes every edge at least once. To solve it, the odd vertices must be connected pairwise with extra edges of minimum total weight. If there are n odd vertices, there are $(n-1)(n-3)\ldots \times 3 \times 1$ possible pairings to consider.

Matchings A **bipartite graph** consists of two sets of vertices X and Y. The edges only join vertices in X to vertices in Y, not vertices within a set. (If there are r vertices in X and s vertices in Y and there is an edge connecting each vertex in X to each vertex in Y then this graph is the **complete bipartite graph** $K_{r,s}$.)

A **matching** is the pairing of some or all of the elements of one set, X, with elements of a second set, Y.

That is, a matching M for a bipartite graph G is a subgraph of G in which no two edges share a vertex. A matching M is **maximal** if it contains the greatest possible number of edges.

If X and Y have the same number of vertices and every member of X is paired with a member of Y the matching is said to be a **complete matching**.

Linear programming The **decision variables (control variables)** are the quantities whose values are to be decided.
The **objective function** is an expression involving the decision variables which is to be maximised or minimised.
The **constraints** are limitations on the decision variables, usually in the form of inequalities.
For a two variable problem you can draw a graph showing the constraints. The **feasible region** is that area of the graph which satisfies all the constraints. The optimal value of the objective function occurs at a vertex of the feasible region (or along a boundary if the constraint and the objective function have the same gradient).

Critical path analysis A project is represented by an **activity network**. The arcs represent the activities. Each node is an **event**, at which some activities finish and other activities can then start. Activity (i, j) is the arc joining event i to event j.
The **total float** $F(i, j)$ of activity (i, j) is defined to be $F(i, j) = l_j - e_i - \text{duration}(i, j)$, where e_i is the earliest time for event i and l_j is the latest time for event j.
If $F(i, j) = 0$, then activity (i, j) is a **critical activity**.

Index

D1

D1